A PROFESSOR, A PRESIDENT, AND A METEOR

A PROFESSOR, A PRESIDENT, AND A METEOR

THE BIRTH OF AMERICAN SCIENCE

CATHRYN J. PRINCE

Prometheus Books
59 John Glenn Drive
Amherst, New York 14228-2119

Published 2011 by Prometheus Books

Inquiries should be addressed to
Prometheus Books
59 John Glenn Drive
Amherst, New York 14228–2119
VOICE: 716–691–0133
FAX: 716–691–0137
WWW.PROMETHEUSBOOKS.COM

15 14 13 12 11   5 4 3 2 1

Library of Congress Cataloging-in-Publication Data

Prince, Cathryn J., 1969–
    A professor, a president, and a meteor : the birth of American science / by Cathryn J. Prince.
        p. cm.
    Includes bibliographical references and index.
    ISBN 978–1–61614–224–7 (cloth : alk. paper)
    1. Silliman, Benjamin, 1779–1864. 2. Scientists—United States—Biography.
3. Science—United States—History—19th century. I. Title.

Q143.S56P75 2010
509.73'09034—dc22

2010029654

Printed in the United States of America on acid-free paper.

For my parents, Marvin and Norma Prince

I did not dream of being favored by an event of this kind in any vicinity, and occurring on a scale truly magnificent.

—Benjamin Silliman

There are more things in Heaven and Earth, Horatio,
Than are dreamt of in your philosophy.

Hamlet, act I, scene 5

CONTENTS

INTRODUCTION

This is the story of a professor, a president, and a meteor.

For most of its first two hundred years, the small town of Weston, Connecticut, bore witness to world events rather than acting as a stage for a world-changing event. Except once. On December 14, 1807, astonished Weston residents watched as a fireball exploded in the skies above. The fireball was actually the first documented meteorite to fall in the then thirty-one-year-old nation. The fall put American science on the international stage and forever changed the way in which Americans, and the world, regarded the celestial ceiling.

That this all came to pass is in no small part due to the work of one man, Benjamin Silliman. Born three years after the signing of the Declaration of Independence, Silliman was the son of an American Revolutionary war hero and a learned mother. A man of faith and science, the twenty-nine-year-old professor graduated from Yale College in 1796. Most Americans no longer know the name Benjamin Silliman, nor that there was a time when American science lagged behind developments in Europe in its institutions and infrastructure.

And yet, Professor Benjamin Silliman shaped modern American science. Certainly Benjamin Franklin, Joseph Priestly, Nathaniel Bowditch, and Robert Hare were all preeminent scientists in their own right. Each one made unmistakable contributions to the field of science. But Silliman did something more. He spoke to the people, connecting with scholars and farmers alike.

After the meteorite fell, Benjamin Silliman scoured the rural town of Weston for data. In doing so he diligently sowed the seeds for what scientists know today regarding the big bang theory, the formation of Earth, and other planetary bodies. Indeed, the Weston Fall was a turning point for American science and for America. The young nation could now occupy a seat at the table with the great European thinkers.

Chapter One

EVERYTHING IS ILLUMINATED

D arkness clung to the early morning sky of December 14, 1807, as Judge Nathan Wheeler set about on his morning stroll. He walked his land every day, finding refuge in the quiet of his Weston, Connecticut, farm.

Suddenly the heavens above Wheeler's farm exploded as a fireball raced across the onyx sky. Everything lit up—his home, his barns, the trees, the blue stone walls.

Looking up toward the northern part of the sky, Judge Wheeler watched the fiery sphere pass behind a cloud, partly obscuring it. Ever observant, the judge thought its edges seemed similar to the sun when covered by the sheerest of mists. The ball rose up from the north and moved in a direction parallel to the horizon. As it traveled, the object slowly climbed toward the west. Judge Wheeler saw it flash with intensity as it passed through clear spots of sky. A shadow of pale light attended the great circle like a train on a wedding dress. Less than a minute later, Wheeler heard three incredible explosions, like an artillery barrage. A succession of rapid reports followed, which he described as similar to a cannonball rolling across the floor.[1] Then the rumble died away, and like the last ember in the hearth, the glow grew more and more until the night sky snuffed it out entirely, leaving only sparks and smoke in its wake.[2]

This abrupt incursion of a luminous object rocked Wheeler's world. Perhaps, just for a moment, he thought a star had come loose and hurled itself across the heavens, bent upon destruction, according to some myth or legend.

The Luiseño Indians of California thought meteors were stars that suddenly moved, and the Eastern Pomo tribe of northern California believed meteors were fire from heaven. According to the Menominee tribe of the Great Lakes, when a star falls from the sky, it leaves a fiery trail. The trace of that trail returns to its point of origin to shine again. The Indians sometimes found the small stars where they had fallen in the grass.

The small town of Weston lies folded into the hills of Connecticut's southeast corner, about forty-five miles from the New York State line. Located less than ten miles from Fairfield and the coast of Long Island Sound, Weston is, as one earlier settler said, "From the sea a day's walke into the country."

As part of the Appalachians known as the Western Connecticut High-lands, Weston wears a slightly craggy face. Steep hills, numerous brooks, and a river or two give the town texture, while isolated pockets of forests stand among the hills and glens. In these remote enclaves, one can envision the moccasin-clad feet of Native Americans who once trod here, before entirely succumbing to the early English settlers.

In 1635, citing religious differences, Roger Ludlow, an Englishman and former deputy governor of the Massachusetts Bay Colony, left the realm of the Pilgrims for "fairer fields." The devout Puritan decided that his future lay southwest toward Connecticut, and so he set off with a band of like-minded people. After scouting out several potential sites, Ludlow settled on the coast of Long Island Sound, in a spot called Uncoway, or "place beyond." When Ludlow and his band arrived, thousands of Native Americans had called Uncoway home for generations. The Pequot Indians didn't welcome the new arrivals. Eventually tensions with the Europeans over trade and territory ruptured into a full-scale war in 1636. Ludlow and his compatriots defeated the Pequots in 1637, in the first of many such wars in the northeast until King Philip's War in 1675, which decimated Native American tribes in New England.

After the Pequot War ended, an uneasy peace returned to the region. The Connecticut General Court, pleased by Roger Ludlow's victory, granted him permission to build a more permanent town on the narrow stretch of coastal land. So in 1639, with teakettles, hatchets, and other accoutrements serving as currency, Ludlow purchased a tract of well-cleared, fertile land from the Paugussett tribe. Ludlow sent word to other colonists from Massachusetts Bay and invited them to live there, to settle and work the loamy soil.

Life in this newly created Eden wouldn't long endure. Religious differences soon arose, and the very issues that provoked Ludlow to set forth from Massachusetts several decades earlier now impelled some recent arrivals to leave. They left the shores of the Long Island Sound to seek their own land north and west of Fairfield. These people wanted to create their own ecclesiastical society. And so they formed a new parish called Norfield. In time, this pocket of land came to be known as Weston.

Southwestern view of the Congregational Church and Academy in Weston, Connecticut. Courtesy of the Connecticut Historical Society, Hartford, Connecticut.

To live in Weston meant choosing a hardscrabble life. Residents were almost wholly severed from the social and religious fabric of Fairfield, a far more established and hence more secure community. Because of this, many Englishmen who settled in Weston in the early 1700s lived there part-time. At night they slept in roughly assembled lean-tos and during the day they tilled the unyielding land. Eventually, traveling back and forth between Fairfield and Weston proved too fatiguing for even the hardiest of these men. They decided to leave behind their friends and families, meetinghouses and schools in order to settle once and for all in Weston. These people became known as "outlivers," for they chose to live in the backcountry, or, as those on the coast called it, the interior.

In 1790, about 2,500 people lived in Weston, one of the poorest towns in all of Connecticut. Eventually the settlement boasted nine grain mills, twelve saw mills, one forge iron manufacturer, four distilleries, four tanneries, three carding machines,[3] three fulling mills,[4] fourteen mercantile stores, three doctors, and one attorney. Self-sufficiency defined the outlivers.

Westonites, like most rural people across the nation, adhered to the cycle of the seasons. Chance occasionally punctuated their strict routine: the chance of sickness and death, the chance of marriage and birth, or the chance of weather and nature. Fevers occurred often, unexpected and violent weather destroyed crops, and on rare instances strange insects arrived in cargo from far-flung ports and decimated carefully planted crops.

Simply arriving in the world had its own perils. One in eight women died in childbirth in the seventeenth century, and one in ten children died before reaching a fifth birthday. Uncertainty and superstition filled the nine months of a woman's pregnancy. People believed children could be disfigured if a pregnant woman gazed on a horrible specter or if a loud noise startled her. Some whispered that an expectant mother's mere glance at the moon might condemn her child to a life of lunacy or insomnia.

Aside from church life, natural cycles conferred order upon the lives of many Westonites. The phases of the moon and the "great celestial wheel of the zodiac" ruled so many lives.[5]

Everyone remained tethered to farm life, even professionals such as attorneys and doctors. Talk revolved around the agricultural year: haying in the summer, repair work in the winter. Though these men and women worked with plants and animals daily, they understood little about the true workings of weather or most things relating to nature. They farmed using practices passed on to them through the generations. Most people behaved according to Old World cultural ideas; many believed astronomical events influenced human affairs. After all, the prevailing culture rooted itself in faith.

Superstitions and old wives' tales swayed townsfolk. For instance, in early America it was said cats that sat with their backs to the fire foretold a cold snap. People learned to read the skies, examining halos around the moon and calculating how quickly the clouds moved. Adages, anecdotes, and rhymes guided lives. So when science finally took hold of this new nation, people began to understand the great unknown of America's physical makeup. Up until then, revolution had been the priority.

Dr. Isaac Bronson sat snugly inside his stagecoach as it rolled along the Post Road from New York City to Connecticut on the morning of December 14, 1807. Bronson looked forward to relaxing in his new country home in Fairfield. He longed to gaze upon the familiar stone walls bordering the property. He loved the coach house and the barn fashioned from the blue stone native to the region. Bronson would wait until spring to see his beloved white dogwood trees bloom.

Born in 1760, Dr. Bronson was the son of a highly respected farmer and his wife, Isaac and Mary Bronson. Shortly before his sixteenth birthday, Bronson apprenticed himself to Dr. Lemuel Hopkins, who practiced medicine in Litchfield and Waterbury. Much of the professional practice involved bloodletting to treat a wide range of illnesses, including fevers.

In 1779, at the age of nineteen, Bronson received a warrant as a junior surgeon, or surgeon's mate, to the Connecticut troops. The young man served under the command of Colonel Elisha Sheldon in the Second Regiment Light Dragoons. Because of Sheldon's frailty, Bronson assumed the senior surgeon's position and performed all the medical duties required for several wartime campaigns. "It is a matter of history that the English officer when acknowledging his identity, requested that an express should be sent to New York for his valet and valise, and that in the interim he requested Colonel Bronson to loan him 'a change of linen.'"[6]

Bronson honed his surgical skills during the war, a grisly residency, to be sure. After being steeped in so much carnage, Bronson traded medicine for commerce once the war ended.[7] He traveled first to Europe and then to India before settling in Philadelphia. After spending some time working as a banker in Philadelphia, Bronson moved to New York City. Growing ever more prosperous, he purchased property on Greenfield Hill in Fairfield in 1796. In 1807, the year of the great Weston Fall, Bronson busied himself establishing a bank in Bridgeport, Connecticut. He and his son Frederic maintained their New York financial and social connections, but they summered in Fairfield.

Deep in thought as he traveled the morning of December 14, Bronson became startled. Suddenly, the inside of his coach blazed. Not knowing if they were under attack, the doctor was instantly revived as the field surgeon during the American Revolution. He urged his driver to make haste. A minute later three terrible explosions sounded, with several smaller ones following suit. Whizzing noises filled the air as rock fragments crashed down on the roof of the horse-drawn coach. Later, he learned the vibrations rattled windows in their casements and shook the very foundations of houses nearly fifty miles away.

Were no one watching, the globe's fiery minuet would have sputtered and died as it fell to Earth, extinguished for all time. Instead, scores of people watched this predawn dance. The extraordinary happening awakened people all along the Hudson River, as far north as Albany. The show lit up the sky over the rolling hills that undulated across the western part of Connecticut and into the neighboring state of New York. The display stunned people as far away as Rutland, Vermont, located more than 220 miles from Weston. It astounded farmers trudging through the snow and frightened bleary-eyed school children staring at their breakfasts. The explosions terrified horses and caught milkmaids off guard.

"I was at the west door of my house on Monday morning, the fourteenth of December 1807, about day light, and perceiving the sky suddenly illuminated, I raised my eyes and beheld a meteor of a circular form, in the southwesterly part of the heavens, rapidly descending to the south, leaving behind it a vivid sparkling train of light," recalled William Page, Esq., in Rutland, Vermont.[8]

> The atmosphere near the south part of the horizon was very hazy, but the passage of the meteor behind the clouds was visible, until it descended below the mountains, about twenty miles south of this place. There were white fleecy clouds scattered about the sky, but none so dense as to obscure the tract of the meteor. I now lament that I did not make more particular observations at the time, and I should probably until this day have considered it to be what is commonly called a 'falling star,' had I not read in the New York papers an account of the explosion of a meteor, and the falling of some meteoric stones near New Haven, Connecticut, which, by recurring to circumstances, then fresh in my recollection, I found to be on the same morning I observed the meteor at Rutland.[9]

Mr. Page eyeballed the fireball to be less than a quarter in size of the moon's diameter. He trembled before its scarlet hue and marveled at its light trail, which seemed to measure roughly eight times the length of its diameter.

In Ridgefield, Samuel G. Goodrich, fourteen years old, woke early the morning of December 14, 1807, to stoke the kitchen fire. Suddenly light filled the room. He looked outside the window and gaped as a brilliant ball of fire nearly the size of the moon scuttered across the sky, northwest to southeast. Then, passing its zenith, it descended toward Earth and, with a trio of explosions, burst into fiery fragments.

"My father, who saw the light and heard the sounds, declared it to be a meteor of extraordinary magnitude," Goodrich said. Writing under the pen name Peter Parley, he left behind some colorful recollections.[10] "It was noticed all over the town, and caused great excitement. On the following day, the news came that huge fragments of stone had fallen in the adjacent town of Weston, some eight or ten miles southeast of Ridgefield. The story spread far and wide, and some of the professors of Yale College came to the place, and examined the fragments of this strange visitor from the skies."[11]

Everyone in the neighborhood heard the stones rush through the air and felt the shock when they struck Earth. One stone, weighing two hundred pounds, smashed against a rock. The rock splintered, its huge fragments plowing the ground for a hundred feet. Later the two professors Benjamin Silliman and James Kingsley took another twenty-five-pound piece back to New Haven. The professors estimated this meteor to be half a mile in diameter and to have traveled through the heavens at a rate of two or three hundred miles a minute.[12]

Some of Goodrich's neighbors believed the folklore about the meteor. One man, a sixty-year-old lieutenant, thought "these phenomena are animals revolving in the orbits of space between the heavenly bodies. Occasionally, one of them comes too near Earth, rushing through our atmosphere with immense velocity, takes fire and explodes!"[13]

Shortly after the Weston Fall, the lieutenant knocked on Goodrich's door. Agitated, he asked to speak with Goodrich's father. The lieutenant entertained the Goodrich family with the most outlandish description.

"In this case of meteors, I suppose them to be covered with some softer substance; for it frequently happens that a jelly-like matter comes down with meteoric stones. This resembles coagulated blood; and thus what is called bloody rain or snow has often fallen over great spaces of country. Now, when the chemists analyze these things—the stones, which I consider the bones; and the jelly, which I consider the fat: and the rain, which I consider the blood—they find them all to consist of the same elements: that is, silex, iron, nickel, &c. None but my animal theory will harmonize all these phenomena, sir."[14]

The lieutenant's depiction was not altogether surprising. During this period, most people continued to seek direction in the heavens. Housewives no longer faced accusations of witchcraft if the butter failed to churn, but no one gave it a second thought if a neighbor planted radishes on a downward angle at the decrease of the moon. This mirrored the plant's tapered shape. No one looked askance at a farmer who planted his crops during the dark of the moon, or at a farmer who believed the optimal time to plant was while the moon waxed. Many people still looked to the moon's cycle to determine the most auspicious time to wean a baby or a calf.[15]

Belief in natural magic lingered in these rural towns like morning dew on the grass. People believed the sun and planets directly affected their earthly lives. As they had during the Middle Ages, many people believed special powers of attraction and repulsion, like a magnet, connected the human body to the universe.[16] Some kept their eyes on the skies in early America, particularly in New England, where several published astronomical almanacs assisted stargazers. Almanacs ranked as the best-selling publications in early nineteenth-century America. Each year, thousands of almanacs sold. Inside, page after page of astrology awaited eager readers. These almanacs contained information such as moon phases, tide tables, planting times, and the setting of religious holidays. Almanacs also had chronologies of world history, poems, and, of course, essays on the workings of the celestial sphere.

The relative isolation of Weston contributed to the poor state of farming. Transporting goods and services between settlements was inefficient and at times dangerous. Therefore the populace, like those in rural towns across the young nation, became bonded to their daily tasks. Westonites made everything, whether they ground flour at the gristmill for bread or felled trees to turn into lumber for homes and barns. The Saugatuck and Aspetuck Rivers powered the mills and yielded fish for food.

Even professionals such as lawyers or schoolteachers farmed small garden plots or several fields. Many households also had woodlots located just outside of town where they harvested lumber and gathered firewood. Most everyone raised pigs, poultry, and cattle. In addition, the town produced maple sugar and syrup, and many settlers kept bees. They concocted tealike drinks from raspberry and blackberry leaves and brewed ersatz coffee from chestnuts and bread. They would rarely buy salt, molasses, rum, tea, or coffee. A barter economy prevailed; one man might lend draft horses to proprietors to pay off a debt, while another man might build a wall to repay a handyman who repaired his barn door. The townspeople relied on each other.

Early nineteenth-century Americans dwelt in a universe defined by its small scale, scarcity, and slowness. People, goods, and information moved like molasses. Farmers happily lived in familiar circles. Yet, news of the nation's affairs did reach rural enclaves. Inhabitants were particularly concerned with commerce and representation in Congress. They paid attention to maritime news, weather, and information about smallpox or yellow fever outbreaks. New England newspapers also noted sermons and cited texts. The *New England Courant* devoted pages to "Speeches, Addresses, Proclamations, and other public Notifications."[17] Yet because eking out an existence sapped the strength of most, improving one's daily circumstances was not a priority.[18]

Just before it pricked a hole through the atmosphere's thin skin, the meteor traveled at between forty and fifty thousand miles per hour. The unbelievable shock of hitting Earth's atmosphere blew its mass into countless fragments. According to eyewitness testimony, the fireball remained visible for an estimated thirty seconds, a considerable amount of time for such a phenomenon to grace the skies.

Many people offered their observations for the historical record. A Mrs. Gardner, whose first name has been lost to time, lived in Wenham, Massachusetts.[19] Heeding Benjamin Franklin's advice of early to bed, early to rise, Gardner routinely woke well before sun-up. Her feet tread across the cold pine boards toward the window of her bedroom, which opened westward for the purpose of observing the weather. This had been her unvarying habit for years. Save for a thin smear of clouds in the west, the sky dawned clear. Peering out her bedroom window, Gardner watched in astonishment as a fiery orb sped boldly across the sky. This intelligent lady observed the event with great attention, corroborating testimony that described the sphere as appearing to be roughly half the size of the full moon.[20]

Mrs. Gardner watched the meteor scamper over the southern part of the barn, which sat squarely in front of her bedroom window. Its well-defined disc resembled the moon so much, Gardner feared for the moon.

"Where was the moon going to?" she asked.[21] Gardner soon realized her mistake. She closely watched the body travel in a direction nearly parallel to the horizon. Then, heading south, it vanished behind a cloud that seemed suspended over her neighbor's farm.

Mrs. Gardner watched the meteor for nearly thirty seconds. At times thin, broken clouds blocked its view from her eyes. Its velocity did not appear to be so great as that of shooting stars, or at least, what she had heard of shooting stars. Unlike some eyewitnesses, she didn't glimpse the light trail but said "its color was more vivid than that of the moon."[22]

The moon did not disappear. Instead, the meteorite roared through the sky, scaring some Connecticut cows. They fled their fields. When farmer Elijah Seeley went to check his cattle he found that they had gone missing. A quick search revealed them shivering in a neighboring pasture. The terrified bovines crossed the other side of that most ubiquitous New England fixture—the rambling blue stone wall. No matter how terrible the weather had ever been on Seeley's farm, his cows had never left home before—where those stone walls, nearly three feet high, surrounded his pastures.

People rose before the sun brightened the morning skies and safely ensconced themselves in their homes shortly before night settled. Most families read the Bible and prayed before meals without benefit of artificial light.[23] Candles were a necessity, but they were used sparingly.

Upon waking, Seeley quickly dressed in his blue- and white-checked woolen shirt, buckskin breeches, white woolen stockings, and double-soled cowhide shoes. The cold morning air bit, so he donned a pair of old stockings, minus the feet, as leggings. Over that, he layered his drugget, a plain cloth vest. Of course, he didn't step outside without his cloth greatcoat and woolen cap. Under the strengthening light of day, Seeley called for his wife. She wore similar attire, though instead of breeches, she wore a plain quilted gown, and midcalf boots of tanned calfskin.

Like Seeley, most Weston households supported themselves through subsistence farming. After bringing the cows home and making sure they were safely in the barn, the couple pushed a wheelbarrow to the fall site. A tumult of meteor fragments lay strewn about the field. They carted away pieces of the black-crusted stones, which greatly differed from the usual crop of rocks otherwise known as New England potatoes.

The hard soil of glacial till—a mix of sand and silt spread over gneissic and granite rock—coupled with the town's hilly terrain made farming hard. Stone walls divided fields, but still "the surface of most of them [fields] was dotted with gathered heaps of stones and rocks, thus clearing spaces for cultivation."[24] As such, Weston cattle were not rotund. The moist, cool climate meant farmers could grow onions, potatoes, corn, oats, rye, and hay.

Just before the incident, Elihu Staples took a purposeful walk about his land. Staples contemplated what he should plant come spring and what fields he would leave fallow. Like many a farmer, he faced a long list of wintertime chores. Wagon wheels needed tending, and the hoes, rakes, and spades would require cleaning and oiling. Barn door hinges always needed a tweak of grease; frequent changes in the weather caused them to creak and groan.

Staples noticed the sky darken. Earth was darker still, the color of a muddy pearl. Suddenly, incredibly loud whizzing and popping noises reverberated through the air. Staples's eyes shot upward. He thought he heard a tornado rip a swath across the tranquil countryside. The fireball skipped across the sky three times, growing dimmer with each successive leap until its final curtain call whisked it off dawn's stage. "It filled all with astonishment and apprehension of some impending catastrophe," Staples recounted.[25]

It sounded like large cannon shot in rapid motion had been fired. "Some compared it to the noise of a wagon, running rapidly down a long and stony hill; or to a volley of musquetry [sic], protracted into what is called, in military language, a running fire. This noise continued about as long as the body was in rising, and died away, apparently in the direction from which the meteor came."[26]

Several miles away from Staples's farm, William Prince and family slumbered in their beds. A great boom abruptly awakened them. They blamed late autumn thunder for so rudely dispatching them from their warm comfort. "Not even a fresh hole made through the turf in the door-yard, about twenty-five feet from the house, led to any conception of the cause, or induce[d] any other enquiry than why a new post hole should have been dug where there was no use for it."[27]

Later that evening, Prince and his sons attended the biannual Weston town hall meeting. Although the townspeople talked excitedly about the event, it was not officially recorded in the meeting's minutes. After the evening meeting adjourned, Prince and his sons hurried home. After hearing numerous reports about stones pounding down all over town, Prince decided to look around his own yard. To his surprise, a hefty frag-ment, weighing approximately thirty-five pounds, lay buried in the ground. Prince broke up the chunk and gave some pieces away as souvenirs.

He gathered garnet, stone, sand, and gravel from the land as did so many other settlers. On occasion, these subsistence farmers sold these gifts from nature to blacksmiths, stonemasons, and lumberjacks to aug-ment their income.

In many ways, life in Weston contrasted sharply with the romantic por-trayal of life in Connecticut. The image of a large home with a blazing hearth, and where "chests of snowy linens" and "well-thumbed Bibles" stood in quiet rooms didn't match the reality of rural life. Early West-onites lived in relative poverty.

Most people lived in small, poorly lit, one or one-and-one-half-story houses. Three to four rooms circled a central chimney.[28] Heavy smells infused the air—many emanating from the barnyard, stable, tannery, tavern, house, hearth, and hard labor. Cleanliness may have been associ-ated with moral purity, but people wore grime like a garment. Woolen clothing worn on unwashed bodies simply smells. Each spring, house-holds scrubbed away the detritus of winter: smoke, soot, bits of bark, pine needles, woodchips, chipped paint, scraps of whittling, knitting, sewing, spinning were all over the place.[29]

Notwithstanding these stark conditions, a dependable familial intimacy governed the lives of early Americans. Together, fathers and sons harvested rye, fixed fences, and seeded fields. Mothers and daughters labored side by side molding candles, cooking meals, and sifting ashes. Farm women gathered together to stitch mattresses and stuff pillows with feathers gathered from migrating geese. Men and women felled trees, chopped firewood, churned butter, slopped hogs, spun thread, and hoed. They endured the necessary backbreaking tasks to keep house and farm from failing in the early 1800s.

Hired men became like family, joining the table during meals. Close quarters ruled; new couples rarely established their own household. Often they lived with extended family and hired hands. The nuclear family was simply the "innermost circle of [a] complex domestic unit."[30] In spite of poverty, people bought and sold land with relative frequency. These business arrangements transcended class, which had more to do with education than wealth. It wasn't until the Industrial Revolution that people promoted the idea of a classless society. In reality, educated men like Isaac Bronson and Nathan Wheeler mingled with men like Elijah Seeley and William Prince. They enjoyed common experiences with nature, culture, religion, and history. All experienced daily life with the same amount of discomfort—no plumbing or central heating. Even people in cities kept domesticated animals such as pigs and chickens.

Townspeople were shaken. It seemed as if the very vault of heaven had been split asunder while invisible hands hurled solid rock to Earth.

Still, it's unlikely anyone, neither Judge Wheeler nor Elijah Seeley, believed that divine intervention had designed the marvel. That kind of thinking might have come into play had cattle fallen ill or crops failed, but most people in the fall zone chalked the event up to weird weather, something the Northeast prided itself on.

It's not that some in Weston, or in New England, didn't think astronomical incidents couldn't influence human affairs. Some, but not all, were slowly shedding long-held associations between astronomy and daily life.[31] While more people made precise observations about their day-to-day lives, most people couldn't believe what they had seen.

These people of Weston, some of whom were learned in the ways of books, while others were learned in the ways of the farm, put the fireball that kited across the sky that chilly December morn down in history as the first documented meteorite to land in America. This sensational event ushered in a new age for the thirty-one-year-old republic and blazed the trail for much of what scientists know today about meteorites and Earth's origins.

Meteorites strike anywhere at anytime. They do not fall only on a few concentrated zones around the world. Scientists map only those stones that have been found. That means scientists and geologists rarely recover meteorites where few people live, such as in hot and barren deserts or in very cold regions such as the Arctic tundra.

Many factors made the Weston Fall remarkable. First, there were eyewitness accounts up and down the Atlantic seaboard. Moreover, the meteor was visible over a two-hundred- to three-hundred-mile area, from New Jersey to Salem in Massachusetts.[32] Fortunately, the fireball of 1807 landed in a place where farmers knew the terrain and were keen at spotting unusual rocks. This is unlike a find such as the Barringer Crater in Arizona, where people discovered the dent in Earth much later. For instance, Berbers in the Sahara constantly find meteorites.

A perfect storm of events allowed the meteorite to fall. Meteorites that successfully fall to the ground are those where Earth's orbit outpaces their own orbit, or where gravity has pulled a meteorite off its original path. If objects are large enough to penetrate deep into the atmosphere, the resulting phenomenon of incandescent gas and dust, called a fireball, can rise to brief, spectacular visual displays and sometimes they can emit noises. An older term, *bolide* (from the Greek *bolis* meaning "missile"), describes large fireballs that explode with loud bangs.[33]

Many early civilizations recognized, or at least believed, that stones fell from the sky. It wasn't until the Renaissance that scientists labeled the idea of stones falling from the heavens as superstition or heresy. Yet in 1807 those people on the ground either thought the world was going to end or that some freak force of nature had visited their small town.

The people of Weston, Connecticut, normally concerned themselves with milking, plowing, and mending.[34] Although the arrival of the Weston meteorite gripped them in a way seldom seen, most townspeople had no means to spread their views much farther than the distance their voices could be heard. That is why a ledger book from 1807 from a local general store lists the same items sold as every other day with no hint that anything special had happened.[35]

However, one scientist did have the means to spread his news and views. In doing so, he would become the first American to report and to describe meteorites, and he would virtually be the sole scientist to do so between 1809 and 1861.

In order for this scientist to hazard an intelligent guess as to the origins and properties of the meteorite's parent body, all the skills and patience of puzzle construction were needed.[36] Fortunately for Weston, the United States, and for the world, one Yale professor embodied that patience. Otherwise this otherworldly event may only have been stitched into the area's oral history. But that's not what happened. The Age of Enlightenment was coming to town in a horse-drawn carriage.

Chapter Two

THE PROFESSORS INVESTIGATE

The carriage loped from side to side like an ungainly camel. Its loose springs forced it to lurch forward on creaking axles. Professors Benjamin Silliman and James Kingsley retained their dignity with difficulty as mud splashed inside without warning. Should the carriage stray off the path, it risked becoming tangled in brambles and briars. Through gaps in the thin curtains the two professors spied New England farmers and merchants trekking through the cold, sledding their goods on horseback or carrying wares on their backs.

As they traveled on, the cultured campus of Yale College in New Haven with its stately brick buildings and homes fell behind. Several miles later, a small rural town came into view.

That Silliman and Kingsley arrived at Weston in a carriage was an event in and of itself, for in 1807, this area of Connecticut looked more like a remnant of the early colonial period preceding the Revolutionary War.

Two days earlier Silliman learned that a great ball of fire had illuminated the sky above Weston before falling to Earth. Upon hearing the news this man of science, Yale's first professor of chemistry, nearly bolted out the door of his New Haven domicile. His writing desk lay open on a stand next to the hearth, and to the right, a high pile of books burdened a small table.[1] Silliman sat correcting proofs for an American edition of William Henry's *Epitome of Experimental Chemistry* when his friend Dr. Isaac Bronson told him about the meteorite. For several weeks, his work on the Henry volume scarcely permitted him time to leave his chamber. He religiously took breakfast and tea in his room.[2] Now a spectacular phenomenon had occurred, and his services were required. So he "broke off every other engagement, and immediately resorted to the scene of this remarkable event."[3] The meteorite truly presented a once-in-a-lifetime opportunity for the young professor.

Silliman would soon speak with eyewitnesses and gain a chance to touch these mysterious stones, freshly arrived from their extraterrestrial descent. The young scientist would be able to investigate the forces binding the rocky matter together.

"I did not dream of being favored by an event of this kind in any vicinity, and occurring on a scale truly magnificent," Silliman wrote in his slim black leather journal.[4] Benjamin Silliman never suspected that divine intervention had hurled the meteorite to Earth, and he had a hunch that it didn't journey from the moon. He knew and often wrote about the numerous reports of meteorites, even those dating back to the earliest periods of recorded history. Yet clear understanding of these marvels eluded scientists since no scientist had yet studied or interviewed eyewitnesses to such an event. So far everything written about meteorites was based on hearsay.

Jeremiah Day, a Congregational minister and former president of Yale College, worked with Silliman. He, too, studied meteors and developed his own theories. Day wrote "A View of the Theories Which Have Been Proposed, to Explain the Origin of Meteoric Stones." Some of those theories were that meteors came from lunar volcanoes; other theories were that they had formed inside clouds. Although Day's work posed different theories than Silliman, the two worked together nonetheless. Silliman himself had read many ancient theories concerning meteorites.

> But, till within a very few years, the subject has been considered, as belonging, rather to fiction and poetry, than to sober philosophy. Men of science, aware, it would seem, of the difficulty of assigning a plausible reason for the descent of masses of stone from the atmosphere, have chosen to intimate, by a significant silence, their disbelief of the accounts which had been given of their fall.[5]

Yet until Silliman and Kingsley came to Weston, people hadn't the means, or the methods, to understand what comprised meteorites or from where these stony intruders hailed. Silliman understood that an event of this magnitude required careful analysis, and he knew he needed a partner to ensure that no stone would be left unturned—literally. So he invited his colleague Professor James Kingsley to accompany him on his mission.

Kingsley, a classical scholar and an expert on ancient languages, joined Silliman because of his skills in interviewing and in writing.

Benjamin Silliman (left) and James Kingsley (right). Courtesy of Yale University Manuscripts and Archives Division.

Benjamin Silliman also chose Kingsley because he shared Silliman's strong religious ethos. James Luce Kingsley was born on August 28, 1778, and his ancestors could be traced to England. They immigrated to Massachusetts sometime before 1630. Historical records indicate his ancestor John Kingsley was one of seven men who established the first church in Dorchester, Massachusetts, on August 23, 1636. His family background explains how over the course of his tenure at Yale, James Kingsley earned a reputation as being one of the most conservative men to join the Yale faculty. So conservative was Kingsley that he disapproved of teaching or studying any book on the Sabbath, the New Testament included. Yet, Silliman's curiosity and his own innate inquisitiveness convinced him to come to Weston, despite their upcoming travel coinciding with the holy day.

The two planned to take a stagecoach from New Haven to Weston, a nearly thirty-mile journey. Silliman gathered his belongings, including compasses, scales, and paper. Then, pausing for a moment before they departed, Silliman jotted a quick letter to his mother, Mary Silliman, on December 18, 1807. "As I am preparing tomorrow morning to investigate the phenomenon, which last Monday morning—of stones falling from the skies. I am in possession of some of them. The fall was accompanied by a meteor and an explosion and loud report in among them were masses of 35 feet even 70 pounds weight. I hope to lodge tomorrow night with brother John to return here about Wednesday."[6] Though Silliman had yet to reach Weston, specimens had already reached the professor. His friend Dr. Isaac Bronson had sent samples along with his letter informing him of the event.

Early on the morning of December 19, Silliman and Kingsley left New Haven for the Fairfield home of David Judson, a Silliman family friend. Judson's house was about twenty-five miles from New Haven and about five miles from Weston. In those years, New Englanders commonly provided overnight accommodations for travelers, whether for dear or distant friends.[7]

As the pair bumped along in their horse-drawn carriage, they noticed how sorely the roads needed improvement. Connecticut roads were notoriously hazardous. It quickly became apparent why Weston was rated with the worst roads in the Nutmeg State.[8] Rutted paths cut through forests and fields, and small streams crossed glorified trails littered with stumps and punctuated with boulders. Travelers discovered that seldom-used trails had quickly sprouted a second growth of trees. Surveyors seemingly plotted roads without considering the area's actual terrain. Bostonians could sail more easily to London than travel overland to Georgia.[9]

Aside from the poor roads, the town's bridges needed attention. Roughly hewn logs spanned the town's numerous brooks and rivers. High waters in the spring and ice in the winter quickly weathered the simple structures; even animals crossed with trepidation.

Of course, tough travel wasn't unique to Weston in the early 1800s. The horse remained the fastest mode of transportation and vehicle for communication. The country's vastness made it hard to traverse even relatively short distances, whether between Wethersfield and Weston, or between New Haven and Norwalk. One stagecoach traveled through Connecticut on the Post Road daily between Boston and New York. Inside sat Silliman and Kingsley, two men on the verge of completely changing what people understood to be true.

As late afternoon wandered into evening, Silliman and Kingsley arrived in Weston. It was Saturday, December 19, 1807. The twenty-six-year-old dark-haired professor leapt out of the carriage, followed by Kingsley. The plush night sky spread overhead, while the moon shone like the eye of a giant cat. Cold perfumed the air. The professors had timed their arrival for the commencement of the Sabbath. Whereas most of Europe had tolerated Sunday travel for the past thirty years, Massachusetts and Connecticut still forbade the practice. Yet, slowly, not only travelers but also innkeepers and others quietly accepted Sunday journeys. Soon the practice would be freed from legal constraints altogether. Nonetheless, being strict in their faith, as were those with whom they lodged, Silliman and Kingsley refrained from delving into their work until Monday.

Upon retiring that night, Silliman felt an incredible sense of urgency to start the task at hand. News of the professor's undertaking had already worked its way into area newspapers; the race for credit and fame had begun. The *Connecticut Herald* was one of the first papers to carry the story.

> A meteor or fire ball, passing from a northern point, disploded [*sic*] over the western part of this state, with a tremendous report. At the same time, several pieces of stony substance, fell to the earth in Fairfield County. . . . Fortunately, the facts respecting this wonderful phenomenon, are capable of being ascertained and verified with precision, and an investigation will, we understand, be immediately commenced for the purpose.[10]

Several days later, after Silliman and Kingsley had journeyed to Weston, more newspapers began retelling the story. On Christmas Eve, people were regaled with the tale of the bright, shining orb in the north. The *Connecticut Journal* published the story on December 24, 1807:

> Fortunately the facts respecting this wonderful phenomenon, are capable of being ascertained and verified by precision, and an investigation will, we understand, be commenced for the purpose. We request gentlemen who may have observed it in distant parts of the State, to favor the public with their observations. It is desirable to ascertain the course or direction of the meteor; the point of compass in which it appeared at different places; the general appearance and velocity; the manner of its explosion, and the time between the explosion and the report.[11]

In fact, even Dr. Isaac Bronson, the local physician, convinced the *New York Commercial Advertiser* to publish his own early account of the fireball. The newspaper recognized a good story and knew its readers would pay to read it. After the Weston Fall, Bronson served as an intermediary between the people of Weston and the learned community of Yale. He brought Silliman and Kingsley to various fall sites and introduced the pair to townspeople who had seen the event.

In those years citizen correspondents like Drs. Silliman and Bronson provided most news reports; teams of reporters were nonexistent. Of course, newspapers carried news items that were really no more than reprints from other newspapers all bundled and repackaged. As news editors unabashedly ripped the headlines from each other, they took advantage of Silliman's presence and his interest in the meteor and circulated it.

Silliman understood the usefulness of the press. He appreciated how it could disseminate information, particularly in regard to science, among laypeople. He recalled how in the latter half of the 1700s the *Pennsylvania Gazette* popularized Benjamin Franklin's many discoveries.[12] He also knew that newspapers could allow men a means to further their personal and professional goals. Benjamin Franklin's *Gazette* provided the model for a newspaper that could present science as "rational, empirical, commercially viable and opposed to superstition."[13] In some way, Silliman knew science could be entertaining and informative. Hence, during the early 1800s, people's exposure to some forms of science had expanded, albeit in very modest measures.

Newspapers generally reflected the increasingly frenetic quality, and pace, of life in America. Although the American Revolution had ended nearly a generation earlier, problems were again brewing with Great Britain and France. It was a time of expansion and change; the nation grew territorially. Aside from the 1803 Louisiana Purchase, people in states such as Connecticut had begun to seek available farming land in places such as Ohio. It was an era of political passion, and the newspapers reflected that too.[14] Adding to this mix was the news that stones had fallen from the sky.

The Weston Fall aroused great curiosity in some and provoked great consternation in others. Yet, most people were more concerned with their crops and livestock and whether they'd have sufficient food stores for the approaching winter. People turned to papers to learn what prices their corn or hay might command, or when a certain ship, carrying essential items, would arrive in a nearby port. Maybe they were selling land or were looking to buy ironware or furniture. Benjamin Silliman's account of the Weston Fall would encourage people to look beyond their daily routine.

A desire to end the debate over where meteorites came from and lead the public from darkness fueled Silliman's pressing need to publicize the Weston Fall. No doubt a certain degree of vanity was involved, but a greater sense of purpose drove Silliman—his idea that shared knowledge would elevate the nation.

Shortly after breakfast on Monday, December 20, Professors Silliman and Kingsley set out to interview witnesses of the Weston Fall. Indeed this interruption in editing and teaching came as a welcome challenge for these men of academe. Silliman began his mission armed with a lot of useful knowledge, including a year's worth of experience lecturing students, time that he used to explore and deepen his understanding of geology. During the summer months the curious scientist had spent countless hours probing the rocks hidden beneath the plains and hills clustered around New Haven. He favored traveling on horseback to these outlying areas, often enjoying the company of friends like Noah Webster, author of the *American Dictionary of the English Language*, as well as newspaper publisher and editor. As Silliman and Kingsley harvested stories and samples about the town, this knowledge served them well. Silliman credited his time in Europe studying to his understanding, and eventual command, of contemporary scientific knowledge. "Had I remained at home I should probably never have reached a high standard of attainment in geology, nor given whatever impulse has emanated from New Haven as one of the centers of scientific labor and influence."[15]

Now, standing in a field in Weston, Connecticut, on a cold winter morning, he realized the incredible scientific challenge that awaited him and his students at Yale. Come autumn the young professor would once again stand before a room of fresh faces, tackling the still novel courses on chemistry and natural philosophy. His lectures would be filled with experiments and walking tours to the countryside, and their content was immensely different from any classroom experience these pupils had ever was experienced. Silliman outlined the relationship between chemistry and geology using the traditional Socratic method. He also amazed his students with personal tales and practical experiments. Few of his pupils were familiar with science, and, in the beginning, even fewer were convinced that science held any value as it pertained to their daily lives. What he learned in Weston would transfer to the lectern at Yale. Silliman, married to precision, was eager to methodically enter data as it was harvested. As one of Britain's well-known men of science, Francis Bacon preached that one had to personally gather as many facts as possible to make certain that what one learned was indisputable.[16]

> Just as if some kingdom or state were to direct its counsels and affairs not by letters and reports from ambassadors and trustworthy messengers, but by the gossip of the streets; such exactly is the system of management introduced into philosophy with relation to experience. Nothing duly investigated, nothing verified, nothing counted, weighed, or measured, is to be found in natural history; and what in observation is loose and vague, is in information deceptive and treacherous.[17]

Bacon's ideas filled a toolbox of techniques that Silliman drew upon to decipher the physical world. This budding American scientist would record eyewitness accounts and toss aside all superstitious stories and old wives' tales. Silliman would admit only the most reliable sources into the scientific record.[18]

With all this in mind, Benjamin Silliman and James Kingsley spent two days visiting several impact sites in the Weston area. They scrupulously separated what they knew to be true from what they could only surmise. In essence they became scientific sleuths.

Using charts, pens, papers, and collecting tools, the pair painstakingly mapped the region. Townspeople led them to the different fall sites. Both Silliman's good friend Dr. Isaac Bronson and an acquaintance, the Reverend Dr. Horace Holley, started a preliminary investigation of the fall area. Unfortunately, their work would be complicated since some curious townspeople had already compromised the scene of the event.

The majority of fragments fell at the primary site that lies in present-day Easton, Connecticut, which borders Weston.[19] The fragments landed about three miles north of a road intersection at the historic 1715 Burton House. The largest fragment weighed slightly more than thirty-five pounds and was plucked from a shallow pit at this site on property formerly owned by William Prince. Today, no trace of the original fall site remains.

Several miles west of Sturbridge Lane, another stone slammed into the soil and sent glacial boulders flying. Another single large fragment, also weighing about thirty-five pounds, was embedded at the foot of Tashua Hill. At Hoyden Hill in Fairfield, on the farm of Elijah Seeley—the farmer whose cows had fled to a neighboring pasture—a small crater measuring roughly 1.5 meters wide and 1 meter deep had been carved into the bedrock. The resident stones had reportedly been thrown and scattered a couple hundred feet away.

Each site yielded precious specimens, and the townspeople provided valuable testimony. Chunks of the Weston meteor survived because the mass broke apart into several smaller pieces as it plummeted to the ground. Individual pieces had a better chance of surviving a violent descent than would a single mass. The atmosphere cannot easily slow smaller stones; therefore they endure far less heating and ablation than do larger masses. These stones land in a more pristine condition.[20]

Except for the largest of meteors, most strike Earth's surface at a fraction of their pre-atmospheric speed. In other words, they crash into the ground at speeds between two hundred and four hundred miles per hour. This means scientists often find newly fallen meteorites littering the surface of Earth like sprinkles on ice cream. Or sometimes they are found burrowed in holes slightly larger than the stones themselves.

Only meteoroids weighing more than 350 tons can penetrate Earth's atmosphere at close to their original speed. These are the behemoths that explode on impact and create immense craters. These are the rare meteors that can radically shift climate and affect animal populations, as did the meteorite that reportedly contributed to the extinction of the dinosaurs sixty-five million years ago.[21]

Although an academic, Benjamin Silliman felt quite comfortable among the country folk as he interviewed them. After all, his father, Gold Selleck Silliman, had been a farmer before he donned the uniform of a Continental soldier. Class distinctions were of little importance to Silliman. Whether a farmer, a banker, or a teacher, all were equal in the eyes of science. Furthermore, Weston and the surrounding towns were linked through blood and marriage. So while neither Silliman nor Kingsley necessarily knew everyone they interviewed, they were in some way connected.[22]

The meteorite investigators interviewed Judge Nathan Wheeler first. Of all the people the pair questioned, Wheeler presented some of the most exhaustive and vivid testimony because, in Silliman and Kingsley's words, he wasn't "influenced by fear or imagination."

Wheeler told them about the light he had seen flash across the sky that December morning. Many people witnessed the light not only in Weston but even hundreds of miles from where the stones ultimately came to rest. However, few described it in such stunning detail as Wheeler.

Most of the people who had the good fortune to witness the Weston meteorite hadn't heard the accompanying symphony of whizzes, pops, and booms. Only those living within a thirty-mile radius were treated to several loud reports as the pressure waves passed, seconds or even minutes after the fireball vanished.

During their stay in the town, the investigative team encountered many people who cleaved to the idea of alchemy and wizardry. William Prince, a wealthy landowner, was one such fellow. Prince and his family barely noticed the explosions. It was only the fortunate coincidence of the town's annual meeting that evening that led Prince to seriously scour his yard. Minutes recorded from the town meeting actually make no mention of the strike, yet the men surely talked about it since such meetings, like church, gave men the occasion to gossip.

During the meeting, they heard that "stones had fallen that morning in other parts of the town."[23] As soon as the meeting adjourned, Prince and his sons returned home as quickly as their horses could ride. They searched the icy earth and from it pulled forth what Silliman described as a "noble specimen." A description: "It was two feet from the surface, the hole was about twelve inches in diameter, and as Earth was soft and nearly free from stones, the mass had sustained little injury, only a few small fragments having been detached by the shock."[24]

The three Princes promptly plunged the hunk of rock into a scalding, bubbling crucible. They hoped to melt out the gold and silver they believed to be locked inside. When that failed, they decided to hack the piece apart, figuring if one stone had value, then many stones had to be worth more.

When William Prince spoke with Professor Silliman, he described the event as sounding like a heavy body falling outside a house. If truth be told, they "formed various unsatisfactory conjectures concerning the cause—nor did even a fresh hole made through the turf in the door-yard, about twenty-five feet from the house, lead to any conception of the cause, or induce any other enquiry than why a new post hole should have been dug where there was no use for it," wrote Silliman.[25] "So far were this family from conceiving of the possibility of such an event as stones falling from the clouds."[26] Throughout the day and well into the dusk of the next, Silliman and Kingsley continued their quest for rocks and witnesses. Half a mile northeast of Prince's farm, the duo finally found another stone, still imprisoned in Earth. This specimen weighed about thirteen pounds. "Having fallen in a ploughed field, without coming into contact with a rock, it was broken only into two principal pieces, one of which, possessing all the characters of the stone in a remarkable degree, we purchased; for it had now become an article of sale."[27]

During their investigation the researchers retrieved many fragments from the lumpy soil. As they scooped up the samples, the two professors noticed how each of the stones appeared different from the others; all of them were oddly shaped, and all wore a thin black crust.

On their second day of fieldwork, Silliman and his partner Kingsley were relieved to discover that although much of the meteorite was rendered unrecoverable because it had been "shivered to pieces," or pulverized into a grainy powder, there were still ample specimens left to analyze.[28] Silliman wasted no time testing each stone he gathered in order to reveal the "exact proportions as well as the nature of the ingredients."[29]

In his report Silliman carefully noted not only the geological ingredients but also the size and shape of the stones. "They possess every variety of form which might be supposed to arise from fracture with violent force," he wrote.

> On many of them, and chiefly on the large specimens, may be distinctly perceived portions of the external part of the meteor. It is everywhere covered with a thick black crust, destitute of splendor, and bounded by portions of the large irregular curve, which seems to have enclosed the meteoric mass: This curve is far from being uniform. It is sometimes depressed with concavities, such as might be produced by pressing a soft and yielding substance. The surface of the crust feels harsh, like the prepared fish skin, or shagreen, a form of rawhide. It gives sparks with the steel. There are certain portions of the stone covered with the black crust, which appear not to have formed a part of the outside of the meteor, but to have received this coating in the interior parts, in the consequence of fissures or cracks, produced probably by the intense heat, to which the body seems to have been subjected. These portions are very uneven, being full of little protuberances. The specific gravity of the stone is 3.6.[30]

Intimately acquainted with their land, farmers are known to spot unusual rocks; thus many meteorites are recovered on farms. After the initial rush in Weston, another mass wasn't discovered for six days. This one lay nearly half a mile northwest from the William Prince homestead. Silliman and Kingsley found the mass in a yawning hole in Earth. It reportedly weighed between seven and ten pounds.

Adding these recovered meteorite fragments to his growing collection helped Benjamin Silliman further understand nature and geology.[31] Even at the start of the 1800s, many common minerals, such as quartz, feldspar, and granite, were just beginning to be identified and named. Silliman and Kingsley's foray into the nation's interior reflected the growing appetite among Americans to categorize the flora, fauna, and fossils around them.

To the delight of Benjamin Silliman, word of this spectacle quickly spread to his mother. Always supportive of her son, Mary Silliman sent a letter from New Haven to Benjamin in Weston. "My dear Son, I want to know the results of your late investigation of that most 'wonderful phenomenon' you mention. The light and report I suppose of the same was seen and heard here."[32]

On December 27, Silliman briefly paused his investigation to answer his mother. As always he kept her apprised of his latest activities, but he also welcomed the chance to unburden himself. His letter underscored the incredible amount of academic pressure he felt to publish his findings quickly, and his fear that when he did, it wouldn't be properly understood. He mentioned his determination to turn out his own article for the *Connecticut Herald*. Yet, he didn't want to rush to publish and risk omitting a salient point.[33]

My dear mother . . . I am at this moment much hurried in finishing the account of the meteor and its consequences. [which will/crossed out] The dissertation will appear in our Connecticut Herald of this week and to that I refer you for ample information. —I hope the account will give

you pleasure. —I will write you again before I set out to Philadelphia—
which will probably be within about two weeks.[34]

Mary, his mother, remained keen on learning the latest: "You will see
in the newspaper of today, a marvelous account of the fall of stones from
the heavens in Weston."[35] Mary Silliman didn't actually read her son's
article in the newspaper, but she told him she had heard that Dr. Isaac
Bronson had sent a letter and published a preliminary report about the
event.[36]

Mary Silliman relished her son's rising reputation. Surely her letter of
January 1, 1808, from Middletown attests to her maternal pride:

> My dear son, I . . . am thankful that I live to address you this morning,
> and to thank you for yours. . . . I go on to thank you for the outlines of
> your journey to Weston and home again since which . . . mentioned the
> paper containing a particular account of this wonderful phenomena you
> and Kingsley have been to investigate. You must have been highly
> entertained. "The works of the Lord are wonderful, sought out by all
> those who have pleasure there in."[37]

On December 31 Isaac Bronson and David Judson wrote to their friend
Benjamin Silliman. They told him several more stones had been found on
Tashua Hill on the farm of Mr. Jennings.[38] Bronson told Jennings he'd purchase the piece for five dollars and then Silliman could rest assured that it
would be sent to "enrich the curious and valuable collection at Yale."[39]

Finally, after two days of painstakingly recording testimony and carefully nesting samples in travel boxes, Professors Silliman and Kingsley
returned to New Haven. They formulated a number of hypotheses. They
required further work in the confines of Yale to confirm or refute their
theories. No matter what, the two professors would contribute to the
growing movement to standardize, rationalize, and map "the natural
world."

Without a doubt the meteor would forever be considered with uncommon interest.[40] Meteorite mania gripped Weston as the investigation continued. The idea that valuable gems could come from space began to captivate farmers who had hitherto lived on scarce resources and paid scant attention to rocks in their backyard other than to clear them for farming.

The Weston Fall became the first cataloged object in Yale's meteorite collection. It remains the oldest and the first scientifically verified meteorite fall in the New World. Its arrival seduced the educated and the uneducated alike, the storyteller and the reporter—accounts of the event appeared in poems, newspapers, and pictures. For once newspapers reported on something other than the political battles between President Thomas Jefferson and the citizens of New England.

Silliman eagerly returned to Yale, where he settled into his basement laboratory. He couldn't wait to analyze the stones and subject the specimens to reagents and testing solutions. He knew he would repeat each step of testing several times. And at the same time he ensured that the accounting would be made public as quickly as possible. As Joseph Priestly, the discoverer of oxygen, once told him, expediting some accounting of an incident, or a discovery, was imperative, even if the research wasn't quite complete. It was the only way to ensure the work would bear Silliman's name.

Now, after days of accruing information, Silliman was equipped to explain what had happened during the predawn hours of December 14. His gathered testimony clearly indicated meteors fell from the sky. Every witness agreed: the falling of the stones happened immediately after the luminous orb flew across the sky.

The work of Benjamin Silliman and James Kingsley had set the stage for the science of meteoritics. American science matured with a citizenry ready to accept and popularize science.

Chapter Three

THE MAN BEHIND THE METEOR

In order to help the New World advance in the province of science, Benjamin Silliman returned to the Old World. Between May 1805 and June 1806, the twenty-five-year-old spent time meeting scientists and studying in England and at the University of Edinburgh in Scotland, where he studied theoretical and analytical chemistry, mineralogy, and zoology.

While in Scotland, Silliman often climbed to the summit of Arthur's Seat, a majestic outcrop that peered down on Edinburgh. Just north of Arthur's Seat jutted Salisbury Craig, a hill that reminded him of a particularly rocky ridge back home in New Haven. One March day in 1806 he climbed Salisbury Craig and experienced a blue sky that stretched above him like a crisply laundered sheet as he crept along the foot of the cliff.

"As it was a very fine morning, I passed several hours in examining the Craig, in pursuit of its minerals," Silliman wrote in his journal. "Soon, however, on looking up, I saw with great consternation a large mass of rock at that instant separating to commence its fall. I leaped over the stones and was sheltered by the friendly column; had I delayed, even the few beats of the pulse while the ruin was beginning its fall, it might have been too late."[1]

Silliman dropped his walking stick and his specimens. He leapt behind a huge rocky column and watched, stunned, as enormous rocks bounded past him down the hill. Suddenly another mass, which weighed several tons, sheered off from the cliff. It thundered down the mountain, filling the air with flying rocks, fragments, and dust. Rock debris covered the path where Silliman had stood seconds before.

"Had the fall been delayed for only one minute, I should have been in the midst of the space which it swept, and a more brief narrative by some other hand would have related the result," Silliman recalled.[2]

In every way Benjamin Silliman took risks and embarked on adventures. He was not content to simply theorize behind a bench in a darkened laboratory. He craved information and endlessly sought answers. Nonetheless, no one acquainted with this young man would have predicted that he would become a scientist of international renown.

Benjamin Silliman was truly a son of the American Revolution. Throughout his early childhood, he heard many stories of wartime Connecticut while being bounced on his father's knee. His father, General Gold Selleck Silliman, served in the Continental Army alongside General George Washington. The elder Silliman, a well-off landowner and lawyer, had enlisted in the army shortly before the war erupted. As a cavalry officer, he commanded the local militia in which he rose to the rank of brigadier general and protected Connecticut's southwest frontier.

Redcoats already occupied Westchester, New York, and Long Island. Now British troops set their sights on the Connecticut coast.[3] The British wanted revenge on the small colony of Connecticut because it fiercely opposed British aggression and the suppression of colonial rights. The redcoats hated this tiny colony and vowed to teach it, and its commander, a lesson. The Continental army couldn't know precisely when the British troops would attack, but General Gold Selleck Silliman did everything to keep the citizens safe.

Throughout the spring of 1779, British warships trolled Long Island Sound near Fairfield, Connecticut. That summer the fight neared the fields and forests of Weston, Connecticut, as British major general William Tryon consumed himself with destroying the morale of the Connecticut people. He and his commander, Sir Henry Clinton, launched a series of hit-and-run raids to convince civilians that surrender was their only option.

Meanwhile, on May 1, 1779, time ran out for General Silliman. Around midnight, armed men crashed through the door of his Fairfield home and snatched him and his eldest son, William, from their beds. Eight armed Tories accompanied the British. The intruders took the general's fusee, a pair of his pistols, his sword, and three hats (one of which was his baby's). Fortunately the raiders missed the general's personal papers. Mary Silliman hid them, along with the family silver, under her nightgown, which the soldiers dared not touch. General Silliman surrendered to protect the lives of his pregnant wife and youngest children.

The captors marched father and son to the shoreline, where a waiting whaleboat ferried the pair across Long Island Sound to a prison in New York. The British captured the elder Silliman because he had been a true thorn in their side. He served alongside General George Washington and enjoyed his confidence, in spite of the number of those in the colonial army who disparaged New England troops. Silliman had fought in the battle of Long Island and served during the colonists' retreat to New York. By all accounts, he acted honorably.

Mary Noyes Fish Silliman worried over her husband's fate; she knew ghastly prison ships bobbed on the water, swallowing the unfortunate. However, calm prevailed. She gathered Gold Selleck, her infant son and the captured general's namesake, along with the rest of her family and servants and headed inland to North Stratford, a less perilous area where they hoped to wait out the war.

With General Silliman removed from the coast, the British successfully struck New Haven in early July of 1779. Troops, with loyalist escorts, stormed Yale College. As British boots stomped through the grounds, faculty hurried about the brick walks carting books and other apparatus to safety. Meanwhile, additional British troops sailed toward Fairfield, where they landed on July 7 in the late afternoon.

The Continental army could not withstand the assault. Waves of British redcoats seemed to wash ashore. They marched to Fairfield Center and burned it to the ground. During this offensive, Major General William Tryon destroyed 83 houses, 54 barns, 47 shops and stores, 2 schools, 2 churches, a jail, and a courthouse. After, as General Tryon's forces marched to and from the burning of Danbury, people fought and fled. Weston men defended their land and provided refuge for their fellow colonists. Weston women fled with their children to Devil's Den, a nearby wilderness spot where men made charcoal.

Meanwhile, Mary Silliman spent the summer safe from the guns. Pregnant and worried about her husband's condition, she continued to carry on the business of homemaking. Years later, she told young Benjamin Silliman about his birth on August 8, 1779: "Even when on that dreadful night, when a band of armed soldiers broke upon our dwelling, and took from my arms my dear protector . . . and I never shall forget the memorable Sabbath morning, where at the hour of six your birth was announced."[4]

With no inkling of when, or if, her husband would return, Mary Silliman tried to fill his space. As a woman who had already lost two children, Mary Silliman doted on Benjamin and his older brother, Gold. Mary's first husband, John Noyes, had died when he was thirty-one years old. She also already grieved over two children. The first daughter, Rebecca, died at age six months; the other daughter, Mary, died at age four.[5]

Ultimately, the British freed General Silliman. A prisoner exchange transpired between two fishing boats anchored in the middle of Long Island Sound. Years later, in 1784, the government billed Gold Selleck Silliman for expenses used to support his troops. Unable to furnish proper receipts, he personally covered the expenditures. His family's wealth thus depleted, the Silliman household felt that Gold Silliman not only contributed his fortune but effectively his life to the cause of freedom.

Later, Benjamin Silliman told veterans of the American Revolution that "the interruption of domestic happiness, the exhaustion of public and private wealth, and the immense sacrifice of lives by which our revolution was accomplished, were esteemed a cheap price for the preservation of our ancient privileges, and for the assurance of future security."[6]

And thus, from infancy Gold Silliman Jr. and Benjamin Silliman learned the value of serving one's country. As part of the first generation to come of age after the war, Benjamin had the luxury of pursuing nearly any profession, just as John Adams had envisioned: "I must study Politicks and War that my sons may have liberty to study Mathematicks and Philosophy. My sons ought to study Mathematicks and Philosophy, Geography, natural History, Naval Architecture, navigation, Commerce and Agriculture."[7] Where Silliman's father had studied war, Benjamin Silliman could study mathematics, politics, or science. His father's legacy endured.

Young Benjamin Silliman spent his childhood on Holland Hill, in Fairfield, Connecticut. The house was perched on a hill above Long Island Sound. Two miles due south of Holland Hill lies the main body of the sound. The broad view nourished his desire to venture forth and learn about the natural world. "Living in a situation perfectly rural, on elevated ground overlooking the country for many leagues . . . in such a situation, we had only to open our eyes in a clear atmosphere to be charmed with the scenery of this beautiful world," wrote Silliman. "A love of natural scenery thus took early possession of our young minds."[8]

The Silliman family traced its origins to Switzerland via Italy. According to Silliman family history, Claudio Sillimandi fled Lucca, Italy, in 1517 to escape religious persecution. He settled in Switzerland, first in Berne and then in Geneva. Eventually his descendants traveled to Holland and England. A Daniel Silliman left England with other Puritan émigrés for political reasons. He settled in Fairfield, Connecticut, in 1658.

Benjamin Silliman's maternal side traced its roots to several pilgrims who had sailed on the *Mayflower*. Mary Silliman was the daughter of the Reverend Joseph Fish and the great-great-granddaughter of John and Priscilla Alden. Voyagers on the *Mayflower*, the Aldens settled in Plymouth Bay Colony in 1620. Mary Silliman came of age in an enlightened home in which she enjoyed the best education girls of her time could expect. Although she was schooled at home, she was learned in art, music, needlework, reading and writing, and some religion and history.

Benjamin and Gold Selleck Jr. enjoyed their father's presence for only a short time. Although he survived the war, he died when the boys were eleven and thirteen, respectively. Benjamin Silliman remembered his father as a sturdily built man with exceptional posture. He recalled his father's hazel eyes and dark hair and how he often smoked a long-stemmed pipe. They remembered their father as a natural-born storyteller, generous and loving.

More than any memento or any keepsake from their father, Mary Silliman wanted the boys to inherit Gold Selleck Silliman's dreams. Therefore, from their earliest days, education occupied a prominent position in the boys' lives. Both parents believed formal education was a civilizing and socially useful process. They believed the Republic required learned citizens if it were to remain stable. Mary Silliman ensured that the boys learned to read and write. Deeply religious, she often asked the children to recite prayers and hymns while she combed their hair and helped them dress. She constantly guided their progress in reading the Bible and other religious books.

Benjamin Silliman grew tall and handsome. Dark eyes and dark hair framed his serious countenance. He displayed manners gracious beyond his years as a result of meticulous grooming.

A view of the buildings of Yale College.

In 1792, following the Silliman tradition, Benjamin matriculated at Yale College, where both his father and his paternal grandfather, Ebenezer Silliman, had attended. Silliman was only thirteen years old when he first laid eyes on the school. Even at this young age, he showed signs of maturing into a serious scholar. He quickly earned the sobriquet "Sober Ben." By all accounts, Silliman studied hard during his undergraduate years at Yale, his dark head often bent over his books well into the night. Throughout his years as a student, he dutifully worked toward self-improvement. His many letters to his mother revealed his gift for introspection: Silliman always respected Timothy Dwight and remembered how, as president of Yale College, Dwight constantly warned youth of "the snares spread around them, in life, and, in warning them, against vice, in every form."[9]

Timothy Dwight, a Silliman family friend, had known Benjamin Silliman from boyhood. He accepted the presidency of Yale in June 1795, a

Yale president Timothy Dwight. Courtesy of Yale University Manuscripts and Archives Division.

few weeks after his predecessor, Ezra Stiles, died. Dwight reigned over Yale for twenty-two years. While he wasn't a scholar, Dwight appointed men who were or who would become scholars. When the Silliman brothers entered Yale College, Dwight presided over their education and served as a father figure. "The President having ever . . . taken a parental interest in the welfare of my brother and myself," Silliman wrote.[10]

Of course, trouble sometimes snared the young student. To be sure, he regularly attended chapel for morning and evening prayers as well as the daily sermon. His upbringing wouldn't have permitted him to veer from that routine. Of course, being young and away from home for the first time Benjamin Silliman occasionally slept through morning prayers, but only because he had stayed up late the night before mastering his lessons. Once when Silliman was a freshman, Ezra Stiles had fined him sixpence for kicking a ball in the college yard.

In 1796, Benjamin and his brother, Gold Selleck Silliman, received their college degrees. The pair studied Latin, theology, and philosophy. Toward the end of his college studies, Silliman wrote his mother about his intention to pursue a legal career, like his father and grandfather before him. Like many talented and ambitious young men of his time, Silliman believed working as a barrister would guarantee him honor, fortune, and stability.

The summer after he graduated from Yale, Silliman returned home to Fairfield. He helped his mother care for their farm on Holland Hill. He also tutored in a private school in Wethersfield, Connecticut. Come autumn he and his brother returned to New Haven and apprenticed at the law office of Simeon Baldwin. In 1799, Benjamin Silliman began working as a tutor at Yale while continuing his study of law. In March 1802, Silliman, nineteen, and Gold Selleck, twenty, passed the bar exam. However, Silliman only briefly practiced law. His mentor, Yale president Timothy Dwight, had other plans for him.

Dwight knew Europe surpassed the United States in the spheres of chemistry and geology. This prompted him to incorporate the comparatively new field of natural sciences into Yale's college curriculum. He was one of the first presidents of a New England college to do so, in addition to men such as John McDowell Provost of the University of Pennsylvania. Though a staunch Federalist and Congregationalist, Dwight knew America remained unexplored. He appreciated how geology and chemistry would help the country's growing mercantilism and industrialization.

To achieve this goal of a science program of study at Yale, Dwight needed to increase the number of professorships on campus. At the end of the Revolution, Yale had few professors of science; for the most part, teaching resided mostly in the realm of tutorials. Usually, these tutors were college graduates who planned on entering the ministry but who also had some sketchy information on the laws of nature and logic.[11] Yale had one professor of mathematics and one professor of natural philosophy who infrequently taught chemistry and natural history. There were only about twenty-one such professors in the nation. Unsatisfied with this system, Dwight convinced the college board, called the Corporation, to appoint a professor of chemistry and natural history. The Corporation agreed, so long as Dwight raised the money to fund the position. It wasn't until 1802 that Dwight found the means.

Dwight sought an American to fill the post, despite the fact that an American would be untested and uneducated in the field. Dwight would not consider a European because "with his peculiar habits and prejudices, [he] would not feel and act in unison with us, and that however able he might be in point of science, he would not understand our college system, and might therefore not act in harmony with his colleagues."[12]

Dwight fussed over the degree to which science might intrude on campus and menace its religious instruction. The degree that politics encroached on both science and religion vexed him. He approached Benjamin Silliman, whom he considered to be a satisfactory candidate. Silliman's upbringing ensured that he would respect religious views. In addition, his father's and his family's wartime experiences meant Silliman appreciated America's need to step out from Europe's shadow.

Silliman was never the obvious choice. True, he and a classmate borrowed one of Yale College's telescopes to observe Jupiter and its known moons in 1795. But aside from that, the young graduate never displayed a special interest in any one area of knowledge. Rather, he demonstrated a rather decent aptitude in a smattering of subjects. But after graduating, studying for law exams became his priority.

"I find no propensity in my system stronger than a wish to be a highly respectable and respected in society," Silliman wrote his mother. "I must act in a particular sphere, and that sphere which is assigned me is the Law. . . . In a country like ours this profession is a staircase by which talents and industry will conduct their possessor to the very pinnacle of usefulness and fame. This pinnacle is constantly in my eye. I am not content (as I once thought it best) to walk obscurely along through some sequestered vale of life. . . . This same thirst for respectability influences likewise all my conduct."[13]

In spite of Silliman's stated goal, President Timothy Dwight took his protégé aside for a walk one warm July afternoon. The two strolled a bit before sitting underneath one of the campus's guardian elm trees. Dwight, who had been a close personal friend of the late General Silliman, implored the aspiring lawyer to set aside the scales of justice. He asked the young man to become a chemistry professor. Dwight appealed to Silliman's ego.

"I could not propose to you a course of life and of effort which would promise more usefulness or more reputation," Dwight said. "The profession of law does not need you; it is already full and many eminent men adorn our courts of justice. . . . In the profession which I proffer you there will be no rival here. The field will be all your own. . . . Our country, as regards the physical sciences, is rich in unexplored treasures, and by aiding in their development you will perform an important public service, and connect your name with the rising reputation of our native land."[14]

Dwight believed that through science one could serve the principles of democracy, patriotism, morality, and even God. He perceived Silliman's love for truth and a desire to be useful to his fellow man.[15] Dwight realized that inside the boy dwelt a man who could do much for Yale. Silliman's appointment signaled Dwight's intention to link patriotism with the growing utilitarian spirit of the age. Science was useful knowledge that could protect and promote the nation's interests.

Dwight surprised Silliman, wrote Silliman: "A profession—that of the law—in the study of which I was already far advanced, was to be abandoned, and a new profession was to be acquired, preceded by a course of study and of preparation too, in a direction in which in Connecticut there was no precedent."[16] And so Silliman surprised Dwight. Silliman grabbed the chance to abandon that predetermined path of law and letters. He was ever mindful of the sacrifices of those who paved the way for his chance to learn. On July 6, 1802, the son of the late General Silliman spoke before the Cincinnati Society in Hartford, Connecticut: "We return with mournful pleasure to seasons of darkness, and kneel with gratitude, over the tombs of those who have bled for their country," Silliman told a hushed audience. "We mark the spot which was once a scene of carnage and survey, again and again, the ramparts of war now covered with verdure. . . . [Y]our feelings of joy for the success of our country are almost swallowed up in the strong sympathy which we experience for the sufferings and death of some distinguished individual."[17]

On September 7, 1802, the Corporation established the Professorship of Chemistry and Natural History and appointed Benjamin Silliman, twenty-three years of age, to be its chair. Once he accepted the position, he began the task of learning chemistry. Silliman knew he needed to study intensively, both at home and abroad. At the time, prospective scientists attended either Harvard College or the University of Pennsylvania. Silliman chose Philadelphia, where American science first flowered. Philadelphia hosted many men who dedicated themselves to science and the ideals science represented: logic and reason.

Shortly after accepting the offer, Silliman again consulted his mother. Upon hearing her son's travel plans, Mary grew increasingly anxious. The remnants of a recent yellow fever outbreak still clung to the city. Having previously suffered the loss of two children, she couldn't abide the thought of something befalling her surviving sons.

Yellow fever had often plagued Philadelphia, most recently in 1793. The outbreak was the most appalling collective natural disaster to strike an American city. Among the sick were Alexander Hamilton and Aaron Burr. Mary knew the epidemic killed four thousand people and that scores more died in successive, smaller outbreaks. Benjamin Silliman's trip worried her because the fever seemed a permanent resident in the city.

"As to the expediency of your going to Philadelphia you and those about you can best judge," wrote Mary Silliman. "It will be a trial to me to have you so far away, but if for the best I must be easy, knowing that you have the same preserver here as there."[18]

Benjamin Silliman reassured her, explaining how his desire to explore this new field outweighed any health risks. Thus he left for Philadelphia to study chemistry with Professor James Woodhouse. Along the way he passed cattle grazing on the pastures of New Jersey. At times, incredibly muddy roads interrupted the pastoral scenery. Wheels dug trenches that filled with water whenever it rained. Thin leather flaps buttoned to the roof and the coach's sides offered meager shelter from the elements. The constant jolting of the carriage over the rough roads made for a rather rude trip, with dusty views of the passing landscape. People, bags, and parcels stuffed the wagon's compartment.

Just a few short years before, this trip would have been nearly impossible. In 1786, it had taken between four and six days, with weather as a factor, to travel from Boston to New York. Stage lines weren't really operational until after 1790. Between 1790 and 1840, Americans began building thousands of miles of new roads and improving old ones. Between 1790 and 1820 private turnpike companies in the Northeast laid roads as well. By 1830, stagecoach lines cut the trip down to under a day and a half.[19]

Silliman enjoyed the varying landscape and its assorted inhabitants. Most Americans knew a bit more about each other's lives than they did a few years before, but their nation was still a federation of diverse regional cultures and economies.[20]

Silliman discovered a whole new world in Philadelphia. The city of seventy thousand people vastly differed from his New England home. Crowded and noisy, the city was a bustling seaport similar to New York, New Orleans, Baltimore, and Boston. Elegant coaches and cabriolets clattered over the cobblestone streets. Draft horses carted goods from the wharves. In such a cosmopolitan city, the cadence of his Yankee voice didn't attract attention; he blended right in.

As the most advanced city in the nation, this Quaker stronghold impressed the New England son. There was less provincialism, more religious tolerance. Here Benjamin Silliman discovered an intoxicating brew of new ideas.

Silliman saw his very first hospital and the first medical college in the country. Sometimes he visited the botanical garden on the banks of the Schuylkill River, which noted botanist John Bartram had tended since 1729. Throughout the centuries, the garden had attracted many visitors. George Washington twice visited in the summer of 1787, and Thomas Jefferson—a man with a green thumb if there ever was one—often walked among the plants. The garden, which flourished with all sorts of greenery, from aquatic plants to trees, fascinated Silliman. He marveled at the artificial pond and stream that adorned the grounds.

In Philadelphia, Benjamin Franklin was omnipresent. The inventor prepared the city to blossom as a scientific and medical center. Aside from the American Philosophical Society, Franklin had his hand in the Pennsylvania Hospital and the College of Philadelphia with its medical school. Philadelphia was the cultural center of the United States. It boasted thriving literary and political societies. Silliman met and mingled with well-known figures in Philadelphia's social and scientific circles. Science flourished here like no place else in the nation.

"It is said that this city has more gentlemen distinguished by their scientific pursuits; I conversed with several well-informed and intelligent men, but there is a cold dryness of manner and an apparent want of interest in the subjects they discuss that, to my mind, robs conversation of all its charm," wrote Frances Trollope, literary figure of the early 1800s and author of several period travel books.[21]

Trollope remarked how the city's elite joined with the blossoming scientific community to both encourage and patronize their endeavors. To land any amount of funding, scientists and their small societies wooed the wealthy and proved they were focused on utility. Thus, a place like the Franklin Institute, which promoted technology education and manufacturing, attracted funding rather easily. Science slowly began to capture the American imagination as people realized it could benefit daily tasks, from charting maps to speeding up travel.

In contrast to New England, the streets of Philadelphia were in fairly good condition. Handsome but not splendid houses lined the streets that ran north and south. The city on the Schuylkill River hosted artisans, fashion-conscious merchants, and professionals. Class divisions might have remained somewhat hidden, but people knew who resided in the more elegant houses and who rode in carriages. There was a pride of aristocracy without the title.[22] Many of these people imitated their European counterparts from the wearing of fine clothes to the appointment of fine furnishings in their homes. Many of these patrician families also supported science and medicine, partly because they felt it was important, partly to guard Philadelphia's position as America's heart of enlightenment, and partly because it was in vogue—many aristocrats in Europe funded science, and it was a way to make a reputation for oneself.

On the other hand, winning patrons in fields such as botany or zoology—President Jefferson's passions—proved more difficult. They didn't hold the allure of bubbling, boiling experiments.

Aside from taking his appetite to learn with him to Philadelphia, Silliman had tucked a candle box in his valise. Inside lay an array of colorful minerals he hoped to identify and categorize. Adam Seybert, a chemist at the University of Pennsylvania, later helped him classify the samples, which consisted primarily of metallic ores, lead and iron, and "a splendid specimen of irised oxide of iron, from Elba."[23]

On the recommendation of friends, Silliman rented a room in a wedge-shaped house at the southwest corner of Dock and Walnut Streets. Mrs. Smith's Lodging House resembled many other such boarding houses. There, as a paying guest, he could expect food, camaraderie from fellow boarders, and the motherly ministrations of the boarding house-keeper. A very select class of gentlemen favored Mrs. Smith's Lodging House, including Robert Hare, a blossoming chemist.

"There were no outward manifestations of religion in our boarding house. Grace was never said at table nor did I ever hear a prayer in the house . . . but rarely have I met with a circle of gentlemen who surpassed them in courteous manners, in brilliant intelligence, sparkling sallies of wit and pleasantry and cordial greeting both among themselves and with friends and strangers who were occasionally introduced."[24]

The company of his fellow boarders pleased Silliman, but sometimes their conduct surprised him. Each evening, he and the other boarders gathered around the long table for dinner. The other men often drank between two and three glasses of spirits a night. Silliman, on the other hand, was unaccustomed to that, as he confided in his journal. He only sipped the occasional frugal glass of wine. "I do not remember any water drinkers at our table or in the house. . . . Porter and other strong beer were used at the table as a beverage. As Robert H. was a brewer of porter, his was in high request, and indeed it was of an excellent quality."[25]

A friendship grew between Benjamin Silliman and Robert Hare, which always proved collaborative, never competitive. Hare, born on January 17, 1781, studied chemistry at the University of Pennsylvania. Together the students attended the lectures of Benjamin Smith Barton and Caspar Wistar, who spoke on natural history. By day, Silliman attended Dr. James Woodhouse's lectures on chemistry held on the first floor of Surgeons' Hall at the Medical College of the University of Pennsylvania. As a lecturer, Woodhouse failed to enchant Silliman. The Connecticut native found him overbearing and stodgy.

> Our Professor had not the gift of a lucid mind, nor of high reasoning powers, nor a fluent diction. The deficiencies of Woodhouse's courses were, in a considerable degree, made up in a manner, which I could not have anticipated. In his person he was short, with a florid face. He was always dressed with care; generally he wore a blue broadcloth coat with metal buttons; his hair was powdered, and his appearance was gentlemanly.[26]

Fortunately for the friends, Mrs. Smith, whom Silliman considered high-spirited, efficient, and indulgent, allowed them to organize a small laboratory in the cellar kitchen of her boarding house. Each evening, Silliman and Hare retreated to their Walnut Street lab and worked under the faint gleam of gas lamps. Some of their housemates worried about the dangers of their self-styled experiments. They had cause for concern. The pair stored gases in proximity to the living quarters, and every now and then, small explosions shook the pie-shaped abode.

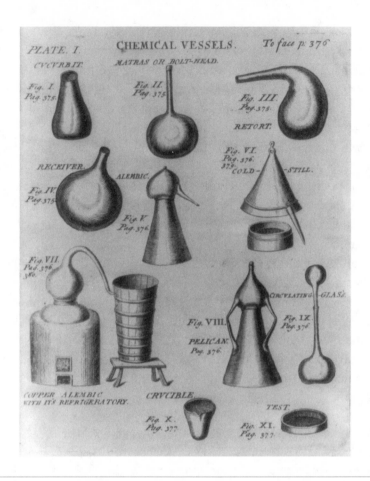

Types of chemical vessels.

Silliman recalled Robert Hare as genial, kind-hearted, and well versed in chemistry. That Hare had progressed in chemistry, still a relatively new field, was remarkable because supplies were difficult to come by. Silliman normally purchased the retorts, glass vessels used for distillation, for a dollar each. Unfortunately, the vessels frequently shattered during experiments, making the research rather costly. However, Silliman didn't mind: "I was rewarded, both for labor and expenses, by the brilliant results of our experiment. . . . I was often surprised, as well as gratified, to find in Mr. Hare an extent of comprehension as well as minuteness of conception and information which made his society a constant scene of entertainment and instruction to me; and in fact, our conversations became so frequent and long on chemistry subjects, that our companions in the house often rallied us on our devotion to this pursuit."[27]

Although Silliman enjoyed allowing for new ideas, he remained fiercely loyal to his religion. Hence, that Professor Woodhouse avoided talk of the Creator during his lectures offended Silliman. "His lectures were quite free from any moral bearing, nor, as I remember, did he ever make use of any of the facts revealed by Chemistry to illustrate the character of the Creator as seen in his works," Silliman said. "At the commencement of the course he treated with levity and ridicule the idea that the visitations of the yellow fever might be visitations of God for the sins of the people."[28]

It's unlikely Silliman believed God wanted to punish the population of Philadelphia, but he didn't condone Woodhouse's oversight of theology. A devout Christian, Silliman wanted religion included in these lectures. He preferred the teaching style of Dr. Joseph Priestly, the Unitarian minister who discovered oxygen, nitrous oxide, and four other gases. Priestley's lectures often combined chemistry with religion.

Not usually eager to court controversy, Silliman nevertheless confronted professors or assignments he perceived as having affronted Christian propriety. For example, on one Sunday Silliman was supposed to visit Charles W. Peale's museum for his American botany course. Peale, a portrait painter and naturalist, started the natural history museum as an extension of his famous portrait gallery. One of his first exhibits was a collection of mastodon bones. The devout Silliman begged to have the day off because it was the Sabbath.

Benjamin Silliman wintered in Philadelphia between November 5, 1803, and March 5, 1804. Throughout his stay he remained a stalwart Federalist of the Washington school. At times the incredible support for President Thomas Jefferson in nearly every scientific and medical institution overwhelmed Silliman. The political and religious divisions between New England and the president were most apparent. All the same, the Connecticut Yankee blossomed in Philadelphia. While there he became a member of Benjamin Franklin's American Philosophical Society. The society enjoyed a reputation as one of the most distinguished and most learned of its kind in the nation. Thus Silliman joined the ranks of many Americans who pioneered careers in science.

In 1804, after completing his studies in Philadelphia, Silliman returned to Yale College. In his first lectures to the senior class, he read an introductory letter on the history, progress, nature, and objects of chemistry.

Silliman held his first lecture in a rented room in Mr. Tuttle's building on Chapel Street. Most classrooms at Yale were sparsely furnished; Silliman's was no exception. His first students sat in a space better suited for storage, devoid of decent laboratory equipment. A fireplace in the classroom fended off the late winter chill. The students listened to the doctrines of latent heat and chemical equivalents. Silliman taught them the forms and laws of crystallization. He instructed them on how to classify minerals, fossils, and rocks in the field. They listened to him expound on the geologic eras, which stretched back into the immeasurable past. He encouraged his students to get out of the classroom and explore New Haven's environs for specimens.

"I subjected whatever came to hand to the action of fire or various chemical agents, and the result was often fortunate in presenting some new discovery . . . thus while I was a teacher, I was still more a learner," wrote Silliman.[29]

Silliman's mentor, Yale president Timothy Dwight, occasionally sat in on one of his classes as an enthusiastic observer. According to Silliman's students, the twenty-four-year-old teacher of chemistry and natural history had a commanding presence, urbanity of address, deep base of knowledge, and a graceful mien. His lectures sparkled; his students' minds crackled with enthusiasm.

Aside from the various teachers and scientists he met, Silliman took the initiative to decipher the art of chemistry. For one, colonial America had no formal guilds where he could apprentice. This had its benefits; when Silliman started his career, the field was a veritable blank slate with a great deal of flexibility in the American system. Silliman sought instruction from many different sources. In 1803, about a year after he joined the Yale faculty, Silliman tracked down Professor John MacLean, a noted chemist, and convinced him to serve as his mentor. MacLean advised Silliman on which pieces of equipment to stock so students could benefit from hands-on learning.

MacLean suggested Silliman acquire a mercurial pneumatic-chemical trough, a galvanic apparatus, a large double convex lens, as well as a simple apparatus for the decomposition of water. MacLean also told the aspiring chemist "an experimenter will every now and then find that he is in want of something, which it was impossible to foresee. However, this is a principle which ought never to be lost sight of, that the more simple an apparatus (provided it be sufficient), the better. A complicated set of machinery, without adding to the accuracy of the experiments, tends to bewilder the student; while a man with plenty of bottles, and tubes and corks may, with the assistance of a blowpipe, vary his apparatus so as to perform an infinity of experiments as well for use as amusement."[30]

Silliman also heeded MacLean's advice about testing knowledge gained. Interestingly, these ideas dated back thousands of years. They actually came from an eighth-century chemist, Jabir ibn Hayyan, who said, "The first essential in chemistry is that thou shouldst perform practical work and conduct experiments, for he who performs not practical work nor makes experiments will never attain to the least degree of mastery. But thou, Oh my son, do thou experiment so that thou mayest acquire knowledge."[31]

To that end, Silliman could often be seen exploring the New Haven countryside armed with a hammer and a blowpipe. Observing and classifying information was essential for him and for other scientists. The seemingly limitless territory of the United States awaited exploration.

Once while still fresh to teaching, Silliman traveled to Boston, another center of science. He wasn't impressed. After visiting Harvard's collection of minerals, Silliman remarked how the gems and stones occupied a lonely spot, gathering dust.[32]

In 1805, Yale College approved the purchase of $10,000 worth of new textbooks and scientific instruments. Silliman volunteered to personally select all the essential supplies on a trip to Europe. In addition to Sillman's regular salary, Yale College allotted him a bookseller's percentage. Silliman successfully convinced the school to both purchase new equipment and allow him to travel. Afterward Yale treasurer James Hillhouse described Silliman as the only gentleman capable of unlocking the college treasury.

However, rather than simply treat the voyage as an all-expenses-paid shopping trip, Silliman prolonged his European trip so he could learn more about the natural sciences. His colleagues from Princeton and Philadelphia penned him letters of introduction to scientists overseas.

Spending time in Great Britain was crucial for Silliman's education since it was still the focal point of scientific study. Britain had textbooks, it had established scientific journals, and it had qualified professors who delivered lectures in centuries-old Oxford and Cambridge. In addition, British scientists enjoyed the established tradition of government support. This piece of information interested Benjamin Silliman because he, too, needed financial support from both the public and the private sectors.

Studying abroad highlighted the utter disparity between Europe and the United States. At home Silliman depended heavily on imported books and journals, a time-consuming venture. If, for example, he ordered books in February, he knew they wouldn't arrive until the following November. It was more productive to personally fetch books and supplies. That prompted his trip abroad.

On April 4, 1805, Benjamin Silliman bade his mother and brother farewell and embarked on a ship out of New York Harbor for London, England. "Tempestuous seas and angry skies" provided a dramatic sendoff as the ship set sail at one o'clock in the afternoon.[33]

An Atlantic crossing was as arduous as it was risky. Everything from boils, seasickness, and fever to thirst, frost, and dampness plagued the anxiety-ridden travelers. Choosing to sail overseas took some degree of courage. But if he were to learn everything he needed to know, he had to cross the ocean. Documenting his every thought helped keep his anxiety and boredom at bay. Silliman kept a diary, written in a graceful but precise hand. He wrote it for his brother Gold as much as for posterity. In it he reflected upon his life thus far, his convictions, and his goals. He also kept careful accounts of more mundane pursuits: "flannel drawers for

$1.43 to keep him warm on trip," "cider, porter, brandy for $14.92." Though misspelled words dot the pages and unformed thoughts are scratched out beyond legibility, Silliman's musings on truth and fact-finding remain quite clear. "To a ship coming in from sea, a false beacon is worse than none," wrote Silliman.[34]

Of course, as befitted a son of New England, Silliman often recorded the weather. Storms swirled about the ship for the better part of the transatlantic voyage. Horizontal sheets of rain became nearly intolerable for the voyager. "The day was dark, stormy and dismal. The ship pitching so much that it was very difficult to stand. The storm increased in violence through the day so that it far exceeded everything which I had hitherto seen . . . night, at length set in, dark and dismal—the tempest raged with more violence than ever, and the fury of the sea was wonderful."[35]

In a letter to his brother, he described the ice he saw near Newfoundland: "I felt a mixed emotion of pleasure and apprehension . . . they were all of a very pure and splendid white . . . few of them were larger than a house or a church, but there were two which might well be dignified with the name of floating mountains."[36]

The journal also reveals Silliman's religious convictions. On the Sabbath of April 14, about 1,200 miles off the coast of Newfoundland, Silliman recorded his views: "The second Sabbath, which we have spent on the water. There is no other attention paid to the day, than that the crew do no other work than merely to navigate the ship . . . instead of public worship and religious retirement, we have rude mirth, and coarse disgusting profanities."[37]

Finally, land ho! Ireland revealed her emerald coast on April 27. A gale kept the ship offshore that first day, but nothing could tamper Silliman's elation upon seeing land after twenty-four days at sea. The ship traveled along the south coast of Britain, passing Stonehenge and the Isle of Wight. At one point, it sailed close enough to the British fleet that Silliman glimpsed seventy ships lying at anchor. He even spied Admiral Nelson's flagship, the HMS *Victory*, at its mooring. When the ship landed at Portsmouth, Silliman watched Nelson embark from the beach in his barge. The crowd roared in admiration, and Nelson waved his hat in a hearty acknowledgment.

Finally, the vessel sailed on to Liverpool. On the morning of May 3, one month after leaving New York, the young man from Connecticut wrote home: "A little before two o'clock we leaped ashore, and realized with no small emotion that we had arrived in England."[38]

Soon after, on June 27, 1805, he received a letter from his mother, Mary.

> My dear Son, my heart glows with gratitude to the presences of men for his allowing you so soon after your departure from your native shores, to announce your safe arrival at your destined port, but you nor I nor any one here I believe dreamt of mountains and islands, but of those dreadful accounts we heard after you went away especially of the loss of the ship . . . with 27 souls that went down with her, among those Mountains of ice gave us great fear that you should share the same fate.[39]

Silliman embraced this new culture from the moment he arrived. While in London he paused and considered the English people and their nation. He gazed upon streets wildly different than those American streets of New Haven and Philadelphia. Chairs, cabs, and carriages crowded the avenues. The London Exchange hummed on the energy generated by the barking of buyers and sellers alike. Silliman visited the Royal Observatory up the Thames, at Greenwich. Not everything pleased him. He recorded unsettling incidents, such as boarding a large slave ship recently arrived from Guinea.

"My own country so nobly jealous of its own liberties stands disgraced in the eyes of mankind and condemned at the bar of Heaven for being at once active in carrying on this monstrous traffic."[40] This experience affected Silliman; he became such a strident abolitionist that he would later supply rifles to Connecticut soldiers fighting in the Civil War.[41]

During his London sojourn, Benjamin Silliman met a practical chemist named Frederick Accum, whose book *System of Theoretical and Practical Chemistry* was one of the first textbooks of general chemistry published in English. Its prose proved easy to read. Silliman attended several of Accum's chemistry and mineralogy lectures held in the professor's home on Old

Compton Street. Accum habitually threw the doors open to the public. Afterward, when he returned to Yale, Silliman emulated this practice, often inviting the general public to learn a few fundamentals of basic geology or chemistry.

Friedrich Accum.

Accum helped Silliman secure instruments and invited the young man to assist on experiments. He coached Silliman in the craft of teaching, offering him ideas on how to interest people with absolutely no scientific background. In addition, Accum introduced Silliman to William Nicholson, editor of the *Philosophical Journal* and one of most prominent English chemists. This contact proved most useful in a few years when Silliman tried to publish overseas.

Then Silliman exchanged London for Scotland, studying there between 1805 and 1806. He took an apartment in a house on Fyfe Street, situated near the old town and the university. In Edinburgh, professors taught science with more sophistication and more polish than in any other city where he stayed. The University of Edinburgh, founded in 1582, had attracted some of Britain's leading scientists, notably Reverend John

Walker. As one of the foremost Enlightenment Age scientists, Walker helped shape the discipline of earth sciences by incorporating the study of chemistry with the study of rocks. Walker had died in 1803, before Silliman arrived. Still, his philosophy endured, and eventually Silliman, too, would blend chemistry with geology.

In Scotland, Silliman had the fortune to study under Thomas Hope, who discovered the temperature of water's maximum density. Hope served Silliman his first taste of geology in a relatively unexplored field for an American. While studying at the University of Edinburgh, Silliman connected with other young American scholars who would also rise to prominence in their respected fields. In Scotland, Silliman shared a house with John Gorham and John Codman of Dorchester, Massachusetts. Codman became a well-respected Congregational minister.

On occasion Silliman closed his books to explore the landscape and discover its earthen secrets. He set off on many solitary hikes and excursions in the area. He climbed limestone cliffs, strolled along riverbeds, and explored castle ruins. The formation of Earth and its rocks enticed the young man.

He also delighted in Scottish history and culture. Three months into his studies, he sent his niece Mary Noyes a lengthy and detailed letter of life in Edinburgh. The letter illustrates Silliman's keen eye for detail and his broad view of the world.

> It is a large & beautiful town; it contains 80,000 people, the houses are of polished free stone; many of them are 10 stories high & some of them 14 & even 16; sometimes there are two stories under ground; people live in them all, from the highest to the lowest. The town stands on several hills, with deep vallies [*sic*] too, so that, as you pass over the bridges, you look down perhaps 100 feet, & see streets, people, houses, & bustle & stir, below you, & as you look up, you see the same on the hills. You can easily conceive that all this must look beautifully at night, when the houses are all lighted up, & the lamps burning in the streets; nothing can be more brilliant; you see illuminated mountains, illuminated vallies, & very high above all the town, on a lofty rock, like the mountains near New Haven, you see Edinburgh Castle, & as you walk in the evening, you hear the French horn, the Bugle Horn, & other martial instruments sounding from the castle, such notes of war, as you my dear, have never

heard, in your peaceful retreat . . . & I hope will never hear. For, this country is at present all in arms, for fear of the French, who you know have for several years, been threatening to invade it. Everywhere, one meets the soldiers clad in red coats, & with shining armour.[42]

Silliman visited tourist attractions such as Queen Mary's apartments, maintained precisely as they were when the royal lady lived there.

In addition, he corresponded with his friend and associate James L. Kingsley. He treasured letters from home. Frequent droughts of mail left him feeling disconnected and depressed. One can appreciate then how rather ordinary news that "the books which you and Mr. Day and Mr. Dwight have written for, are all ordered" cheered him enormously.[43]

The introductions from fellow Americans eased his entrance into Scotland's scholarly society. During his time in Edinburgh he met several prominent European scientists at the Royal Institute, Cambridge University, and the University of Edinburgh. Silliman even met Dr. Robert Darwin, the father of Charles Robert Darwin. Even if they didn't regard him as an equal, the scientists and manufacturers warmly received the young American.

For the most part, Silliman's papers, letters, and journal focus on science and technology. They are a purposeful reportage that he hoped to share with the growing body of scientific and technical professionals in America. These letters represent the degree to which he, and his country, became ever more fascinated by science. At the same time, his letters to family, packed with observations on tourist venues, illustrate his ability to communicate with laypeople. The journal—like Silliman—bridged the gap between experts and generalists, between scholars and sightseers.

Then on May 2, 1806, after nearly twelve months abroad, home beckoned. Yale College required his services. John Gorham, his roommate in Edinburgh, accompanied Silliman. They enjoyed a lasting friendship, remaining pen pals for years. Gorham lived in Cambridge, Massachusetts, and eventually became a chemistry professor at Harvard College.

Silliman credited his time at the University of Edinburgh for molding his professional character. "When I left New Haven, in March 1805, on my way to England, I was quite in the dark regarding the nature of rocks that surrounded me at home, and I have already stated how light broke in upon me in Edinburgh," Silliman wrote. "It is obvious that, had I rested content with the Philadelphia standard, except what I learned from my early friend, Robert Hare, the chemistry at Yale College would have been comparatively an humble affair. In Mineralogy, my opportunities at home had been very limited. As to geology, the science did not exist among us, except in the minds of a very few individuals, and instruction was not attainable in any public institution. . . . Here my mind was enlightened, interested, and excited to efforts, which, through half a century, were sustained and increased."[44]

The return trip across the Atlantic Ocean took several weeks. Silliman landed on the Long Wharf in New Haven on June 1, 1806. Though only a few weeks remained in the summer term, he gave a chemistry course to the Yale senior class. He greatly anticipated incorporating all that he had acquired overseas, the knowledge, books, instruments, and specimens, into his lectures. He transformed his classroom. Glass tubes shimmered, a Nooth's machine for impregnating water with carbonic acid gas lorded over the vials, and bottles and corks adorned the shelves.

Silliman enthusiastically expanded his repertoire. Not only did he deliver chemistry lectures, he began to lecture on geology and mineralogy. That first summer home, he gathered many new rocks and minerals to enrich his lectures.

In September 1810, he read a memoir, *Sketch of the Mineralogy of the Town of New Haven*, before the Connecticut Academy of Arts and Sciences. Founded in 1799, the academy is the third-oldest such society in the United States. Its mission is to disseminate scholarly information not only among scientists but among the general population as well. Silliman impressed the members of the Connecticut Academy with his newfound insight into earth sciences. Finally, chemistry and natural history had attracted the attention of educators. Until now, these fields were considered offshoots of medicine.

Benjamin Silliman might have been a child of the Revolution, but he fast became a father of scientific awakening. A diligent observer of people and objects, Silliman earnestly illustrated and incorporated discoveries into the growing field of natural science.

Yes, Joseph Priestly had shown the existence and properties of oxygen. Joseph Black showed doctrines of latent and specific heat. And, Henry Cavendish had shown the existence of hydrogen and decomposition of water, while Antoine Lavoisier had demonstrated the chemical changes involved in combustion and evaporation. But not one of these scientists truly possessed the ability to transmit his passion to the untrained observer. That quality belonged to Benjamin Silliman alone.

Chapter Four

THE STATE OF SCIENCE

When Benjamin Silliman assumed his post as Yale College's first chemistry professor, he pictured doing more than delivering one lecture after another. He envisioned leading his students in practical, hands-on experiments. However, becoming fully versed in the language of equations and formulas would take time; the subject of chemistry was still very much in its infancy, and science still attracted only a small number of people.

Most of these sailors and merchants who hailed from seafaring states such as Connecticut relied on math, astronomy, and meteorology to plot charts and courses. City planners and explorers surveyed territories and cities using classical astronomy. Many other small cities and towns, including Weston, were laid out from east to west and from north to south using similar techniques. The work was practical. Benjamin Silliman explained the usefulness of astronomy, as well as its pleasures: "Astronomy is not without reason, regarded, by mankind, as the sublimest of the sciences. Its objects, so frequently visible, and therefore familiar, still do not lose their dignity, because they are always remote and inaccessible. . . . It may be mentioned also, without impropriety, that the observation of the heavenly bodies is a rational source of amusement."[1]

Among the best-known astronomers at the time were David Rittenhouse and Nathaniel Bowditch, both of whom were self-educated. Bowditch learned his trade on sailing ships. David Rittenhouse assembled precision clocks and used the proceeds to finance his studies in mathematics and astronomy. A native-born Pennsylvanian, he devised an ingenious instrument called an orrery. The clockwork model allowed users to "see" Earth orbit around the sun and how the moon relates to Earth. A third astronomer, Andrew Ellicott, was also a surveyor and clock maker. Ellicott did some early work on calculating the rise and fall of heavenly bodies. Benjamin Silliman knew their work and considered it while walking across the snow-trampled ground in search of specimens after the Weston Fall. He used the knowledge to calculate distances, the meteorite's height, and its rate of descent.

The New Haven Puritans, similar to the Massachusetts Bay Puritans, needed educated ministers to preserve their religion and their politics. Connecticut also needed a continuous supply of educated laymen to help develop and retain its hold on growing commercial enterprises. In 1701, Elihu Yale had founded Yale College with the mission to educate young men primarily in theology, but with more than a smattering of general knowledge. The school mission statement declares: "Every student shall consider the main end of his study to wit to know God in Jesus Christ and answerably to lead a Godly sober life."[2] The school slowly developed into a place "wherein Youth may be instructed in the Arts and Sciences [and] through the blessing of Almighty God may be fitted for Publick employment both in Church and Civil State."[3]

By the time Benjamin Silliman attended Yale it was clear the dame schools (private elementary schools where students were taught in the teacher's home) and hornbooks of the colonial era no longer had a place. Most communities understood that education made democracy work, and knowledge would make democracy strong.

By 1800 students from many states flocked to New Haven to attend the small college. By 1810 Yale included three separate colleges. In addition, the campus had a chapel with rooms for rudimentary laboratory apparatus, and a library containing seven thousand volumes. The college needed land to expand almost from the beginning. It didn't own the land facing the New Haven green, and, according to Benjamin Silliman, it was "filled with a grotesque group—generally of most undesirable establishments, among which were a barn—a barber's shop—several coarse taverns or boarding houses—a poor house and house of correction—and the public jail with its prison yard—and used alike for criminals—for maniacs and debtors. The wild laughter of the insane resounded in the college yard."[4]

Private schools and colleges like Yale dedicated to mathematics, chemistry, philosophy, and astronomy laid the groundwork for what have come to be known as the Ivy League schools in New England. Aside from the practical arts, religion was an integral part of Yale College's curriculum, just as it played a fundamental part in the life of New Englanders. Yale students attended public prayer upon rising each morning and before retiring each evening. In the first decades, they studied little else except Greek and Latin classics and theology. However, as time passed the institution's elders understood that Yale needed new research outlets and innovative programs to ensure its longevity and legacy. Thus, Yale president Timothy Dwight pursued the new science of chemistry.

Dwight focused on science when he replaced Ezra Stiles as president in 1795. The former preacher perceived science's intrinsic value, that it could serve the wants of man and propel society forward. The new Republic needed answers, and American citizens needed more control over their lives. However, to achieve this, the various states of the new union needed more contact among themselves. Most American cities lacked good communication; without reliable roads and a dependable postal service, towns and villages remained cut off from one another. As a result, American chemists remained utterly dependent on Europeans for their information and research. Thus, the exchange of new scientific ideas in America suffered.[5]

Two hundred years before Benjamin Silliman studied in Edinburgh, men such as Johannes Kepler and René Descartes revolutionized how people looked at the skies. In 1605, Kepler announced, with great fanfare, that he no longer believed an internal soul moved the planets through space: "I am much occupied with the investigation of the physical causes. My aim in this is to show that the machine of the universe is not similar to a divine animated being, but similar to a clock."[6] In 1596 he wrote *Mysterium cosmographicum*, or *The Secret of the Universe*, which calculated the distance between the sun and the six known planets of the solar system as it was known at the time. Kepler's work uncovered how planets passed through space as well as perpetuating the theory that mathematical law governs nature. In 1630 Descartes compared mechanical clock movement with natural bodies such as planets.

"Science must be known by its works," said English philosopher and scientist Francis Bacon. "It is by the witness of works rather than by logic or even observation that truth is revealed and established. It is from this that the improvement of man's lot and the improvement of man's mind are one and the same thing."[7]

Bacon lived during the 1600s and served as lord chancellor in the court of King James. Bacon believed that in-depth study of the seas, the planets, and the world as a whole would alleviate human misery and contribute to the quality of life. In addition, while some "natural philosophers" contentedly theorized from inside the classroom, a growing number realized the import of Bacon's vision—that one needed to actively collect facts and engage in the process of experimentation. This was a starkly different way of considering the world, but nevertheless, it was an idea that had been germinating since Galileo: everything must be submitted to reliable observation and mathematically disciplined reasoning. It was a radical idea because it rebelled against the traditional view of the clergy that heaven could not be scrutinized. According to the dogma of the time, heaven was perfect. These revolutionary Enlightenment ideas inspired the work of men such as David Rittenhouse, a clock maker turned scientist; Nathaniel Bowditch, an astronomer; and, of course, Benjamin Silliman.

Until the start of the nineteenth century, interest in science had merely buzzed in the background. More than two decades after Great Britain and the United States inked the Treaty of Paris, America had little or no scientific standing in the world. Few scientists walked the halls of the nation's schools, and nationwide there were barely twenty-one institutions of any quality. Few academies of medicine existed, and there were none dedicated to agriculture or literature. There were no prizes bestowed on prestigious works. Moreover, the democratic culture permeating early American life meant there existed a sense that individuals should engage in only income-producing activities, without the benefit of patronage. Government support for education, let alone science, was minimal, so budding scientists had to locate and secure private funding. No old boys clubs existed; budding new scholars had to rely on the more established scholars for acceptance. Without letters of introduction, Benjamin Silliman would have found himself on a rather lonely course in Philadelphia and overseas in London and Edinburgh.

Chemistry requires a great deal of glass. The earliest-known use of glass for experiments dates back to the third century when Zosimos of Panopolis, in Upper Egypt, raved about the quality of glass vessels imported from Askalon in Syria.[8] In 1607, the Jamestown colonists' had glassmakers to create glass bottles and beads to use as currency with the Native Americans. However, true glasshouses didn't exist until the mid-1750s. The New York *Gazette* published an advertisement for the New York Glass House Co., located at North River. It sold glass bottles and "other Chymical Glasses made with all expedition."[9] In August 17, 1769, Richard Wisters's Glasshouse advertised in the New York *Journal* or *General Advertiser*: "Bottles, etc., Receivers and Retorts of various sizes also electrifying Globes and Tubes, etc. As the above mentioned glass is of American Manufactory; it is consequently clear of the Duties the Americans so justly complain of, and at present it seems peculiarly the Interest of America to encourage her own Manufactories more especially those upon which Duties have been imposed, for the sole purpose of raising a Revenue."[10]

However, in the early 1800s very few glass chemical instruments were available, and nascent scientists such as Benjamin Silliman often used deficient equipment. While living on Walnut Street in Philadelphia, Silliman purchased decent glass retorts, glassware that could be heated for experimental purposes. Now in New Haven, the professor needed a supplier closer to Yale. Silliman requested that a Mr. Mather—Silliman never mentioned his first name—a glass manufacturer in East Hartford, fashion some glass retorts so he could distill liquids in his laboratory. Mather confidently told Silliman he could fill the order, although he had never before seen a retort. Unfortunately, the manufacturer didn't execute the project quite as planned.

"I had a retort the neck or tube of which was broken off near the ball—but as no portion was missing, and the two parts exactly fitted each other, I sent this retort and its neck in a box, never dreaming that there could be any blunder," Silliman wrote. "In due time, however, my dozen of green glass retorts, of East Hartford manufacture, arrived, carefully boxed and all sound, except that they were all cracked off in the neck exactly where the pattern was fractured; and broken in neck and ball lay in state like decapitated kings in their coffins."[11]

To Silliman this proved the pitiable state of manufacture of chemical glass in Connecticut, if not the entire country. The same blunder probably wouldn't have occurred in Philadelphia or Boston, two cities with more established scientific centers. That's why early scientists such as Benjamin Silliman frequently ventured overseas. Studying and training in Europe meant earning a greater degree of professional sophistication as well as finding exceptional equipment. If one wanted new tools, one traveled to London. One outfit in particular, Fidler and Troughton, made "extremely sensible balances," according to Silliman.[12] Troughton's built its reputation on custom order equipment. Customers expected quality products, but they also expected to wait up to two years for goods to be delivered.[13]

Aside from having trouble locating proper equipment, new scientists had difficulties finding laboratories in which to conduct experiments and research. The German chemist Andreas Libavius (1540–1616) once described the perfect "chemical house," or laboratory. It consisted of several rooms including a prep room, a storeroom, and a room for experiments. In 1631, John Winthrop, the son of the first governor of the Massachusetts Bay Colony, organized the first laboratory in the United States.[14] Little information remains of how Winthrop's laboratory looked. Most likely it was sequestered in a dank basement rather than inhabiting spacious quarters after the fashion of Libavius's ideal chemical house. Yale didn't possess anything remotely resembling a modern laboratory until 1804. Before then a single room served the purpose.

"It [Silliman's lab] was in the old college, second loft, northeast corner, now No. 56," wrote Benjamin Silliman. "It was papered on the walls; the floor was sanded, and the window-shutters were always kept closed except when visitors or students were introduced. There was an air of mystery about the room, and we entered it with awe, increasing to admiration after we had seen something of the apparatus and experiments."[15]

America's colleges also needed books. Most of the country's new schools and societies lacked the necessary volumes. During the first 150 years of settlement in colonial America, few medical books, pamphlets, and broadsides were written, published, and disseminated. There were early attempts to publish useful texts; two Philadelphia men, Benjamin Rush and James Woodhouse, issued *The Young Chemists' Pocket Companion* in 1797. However, according to the 1804 edition of *The Catalogue of All the Books Printed in the United States*, of the 1,338 books published that year only about 20 were scientific in nature.[16]

For many, science was a purely part-time hobby practiced between preaching, medicine, and politics. Most scientists experimented in private, home-based laboratories. Some organized occasional weekend natural history outings but accomplished little academic work. So new was the concept of higher education in science, or most any other subject for that matter, that the study of earth sciences usually took place in medical school classrooms.

In addition, unlike the British and French governments, the US government still hadn't expressed a sustained interest in supporting science, though President Thomas Jefferson had sponsored the Lewis and Clark expedition known as the Corps of Discovery. In May 1804, Meriwether Lewis, William Clark, and their coterie of explorers set off to explore the nation's geography, minerals, flora, and people from the newly purchased Louisiana Territory to where the sun sinks into the Pacific Ocean. Every day across the young country, people discovered animals, plants, and minerals for which they had no names; the expedition was a scientific survey of the nation's present and future. It illustrated President Jefferson's belief that science could and should help government advance and that government should in turn nurture science.

Thomas Jefferson.

The new century ushered in hopes of a new age, and Thomas Jefferson shone as a beacon for those invested in science. He believed his duties included establishing a more firmly democratic republic where sciences occupied a prominent place.

"Science is my passion, politics, my duty," he famously said while serving as secretary of state in 1791.[17] He ardently believed science would help civilization progress both morally and intellectually. He wanted science education to be mandatory in schools. Like Benjamin Franklin, he believed science could help citizens achieve common ground. He also supported the idea of total freedom to investigate and inquire.

There were other amateur scientific accomplishments during these years. In 1801, Pennsylvanian James Finley built the first modern suspension bridge. That same year Charles Wilson Peale and his son Rembrandt Peale excavated and assembled two mastodon skeletal remains in Orange County, New York. The Peales installed one set of the remains in their natural history museum in Philadelphia and sent the other on tour in 1802 to New York and to London.

Alexis de Tocqueville in *Democracy in America* pointed out these deficiencies in the young nation: "There is no denying the fact that among civilized nations today, few have made less progress in the higher sciences or produced a smaller number of great artists, illustrious poets, and celebrated writers than the United States."[18]

During the American Revolution, the British occupation quashed burgeoning scientific work in cities like Boston, Charleston, New Haven, and New York. Natural philosophy, mathematics, and other intellectual pursuits suffered from the dislocations of the war. The Philadelphia-based American Philosophical Society had only recently reemerged from the upheaval. Before the Revolution, this scientific institution, America's first, fared rather well. But when the British occupied Philadelphia, the society scotched its plans to build an observatory platform in the State House yard. It took Benjamin Franklin's 1785 return from France to inject new life into the decades-old society. The observatory finally came to pass when David Rittenhouse became the society's president, succeeding Franklin. A mentor of Jefferson, Rittenhouse erected the observatory near his own home.[19]

Several positive yet unforeseen consequences resulted from the war, aside from liberating the country from Great Britain. Foremost, the war boosted cultural nationalism. This newfound patriotism included recognition that Americans needed to promote the sciences. Americans craved not only independence from Britain but worldwide respect as well. They yearned to show the world their intellectual power. Suddenly, scientific accomplishment became a measure of civic pride.[20]

Patriotism pushed intellectuals like Benjamin Franklin and Thomas Jefferson to promote American science. Jefferson understood that

America needed to lean on Europe to lift itself up and move forward.[21] Only then could it counter the negative attitude Europeans had toward American science. Napoleon Bonaparte disparaged Britain by calling it a "nation of shopkeepers."[22] Some in the United States worried it, too, would be no more than a nation of shopkeepers if it continued in Europe's shadow. Science became the key to loosen the chains that kept the young nation academically dependent on scholarship from overseas; a dependence not borne of affection but of necessity. In the early 1800s most European intellectuals pitied those posted to the United States, considering them exiled into social and political oblivion. There was an "antagonistic/disinterested attitude displayed by England reviews regarding American science. There is a universal desire and endeavor [*sic*] to forget America, and a unanimous Resolution to read nothing, which shall bring it to their thoughts. They cannot recollect it without pain."[23]

America tired of living in Europe's shadow.[24] It had ceased being simply a supplier of raw materials to England under the reign of King George III. In this period, America worked diligently to forge new links with the Continent. A patriotic urge to silence British sneers and quiet Continental critics gripped American artists, businessmen, writers, and even scientists. Independence now had less to do with politics and everything to do with culture and science.

In 1743 Benjamin Franklin said, "The first drudgery of settling new colonies which confines the attention of people to mere necessaries is now pretty well over; and there are more in every province in circumstances that set them at ease and afford leisure to cultivate the finer arts and improve the common stock of knowledge."[25] He founded the American Philosophical Society in May of that year. The organization helped link the colonies since it encouraged scientists from every region to share their theories. Under its direction men of science exchanged papers at least four times a year. Franklin wanted to delve into "newly discovered plants, herbs, trees, roots, their virtues, uses . . . improvements in any branch of mathematics . . . new arts, trades, and manufactures . . . all philosophical experiments that let light into the nature of things."[26] In addition, the society printed essays and news of its projects in regional newspapers. The philosophical society devoted itself to improving the nation.

While most Americans still regarded the colonies as unrelated to each other, in a loosely arranged confederation, Franklin viewed them as belonging to a larger, single organism. His extensive travels from north to south and overseas helped to shape this outlook. "Knowledge is of little use, when confined to mere speculation," Franklin said. "But when speculative truths are reduced to practice; when theories, grounded upon experiments, are applied to the common purposes of life; and when, by these agriculture is improved, trade enlarged, the arts of living made more easy and comfortable, and, of course, the increase and happiness of mankind promoted; knowledge then becomes really useful."[27]

In 1780 John Adams founded the Boston American Academy of Arts and Sciences to investigate areas that would "enrich and aggrandize these confederated States."[28] At the time, Harvard College was receiving the bulk of its funding from Boston's mercantile elite.

Franklin, the bespectacled inventor from Philadelphia, understood the country would only progress if every single citizen, not merely the educated upper class but the ploughman as well, acquired basic knowledge. Franklin's zeal for scientific accomplishment reflected the waves of Enlightenment theory flooding America's shores. He dreamed of a nation in which libraries, hospitals, colleges, and other similar organizations were the measure of its success. He believed these institutions benefited the populace far more than churches and seminaries.[29] Franklin wanted Americans to create a modern, unified nation through the pursuit of common knowledge.

In these years people were divided on the proper course for public knowledge. There were some who championed public education because they deemed it practical and useful. There were those who promoted education because it was classical and aristocratic. Yet those like Franklin and Yale's Dwight began to see that education could serve a purpose and have a function.

The notion of a common knowledge base complemented the idea that one should pursue a warrantable calling, as the puritan minister John Cotton had preached in 1641.[30] Cotton believed one should only pursue those avenues that benefited both individuals and society. Science easily met the criteria. It could help the New World unlock its material promise (a public good) and help build a scientist's reputation (an individual gain). The idea that science could and would support material and economic interests was condoned so long as self-interest and greed left it untainted. In the eyes of the Founding Fathers, useful knowledge equaled common knowledge. "The common herd of philosophers seem to write only for one another. The chemists have filled volumes on the composition of a thousand substances of no sort of importance to the purposes of life. . . . [They should use it for] brewing, making cider, to fermentation and distillation generally."[31]

Soon academics found chemistry useful, in spite of the fact it was still relatively new—it had replaced alchemy only in the last century or so. "Chemistry is now cleared of its rubbish, and stands on the broad basis of demonstration. Modern Chemistry affects no mystery—it appeals to no hypothetical existences—it claims no supernatural aid; but the substances of which it treats, are all actually produced and made obvious to the senses of the most unlearned," Silliman said. "No principle is professed which has not been previously demonstrated by experiments, and no language is used which a good understanding and moderate application will not render intelligible. Indeed the language of Chemistry has been more simplified and reduced to principles more strictly logical, than that of any other science whatever."[32]

In time Yale College embraced chemistry within its curriculum. The field helped reinforce new ways of thinking for students who increasingly fell under the influence of Francis Bacon and other thinkers. As the Scottish philosopher David Hume wrote in *A Treatise of Human Nature*: "Where experiments of this kind are judiciously collected and compared, we may hope to establish on them a science, which will not be inferior in certainty, and will be much superior in utility to any other of human comprehension."[33] Hume believed that when properly applied, science could furnish mankind with both prosperity and freedom.

When it came to chemistry, Thomas Jefferson believed in the utilitarian philosophy. Thomas Cooper, a close friend and sometimes collaborator of Jefferson, shared his outlook. Cooper emigrated from England and eventually became professor of chemistry at Dickinson College in Carlisle, Pennsylvania. Both Jefferson and Cooper considered chemistry a tool to bake better bread, churn smoother butter, and ripen tastier cheese. As the United States shifted from an agrarian economy to a mercantile one, this way of thinking grew more popular.

Later, with the advent of the Industrial Revolution, the lower and working classes began to contemplate science. Some understood that science could improve their lot and their social class. It coincided with Jefferson's belief that mankind should share all useful improvements without restraints: "He who receives an idea from me, receives instruction himself without lessening mine; as he who lights his taper at mine, receives light without darkening me."[34]

Heretofore, science had developed regionally throughout the country. Unlike Europe, which had Paris and London, the United States couldn't point to any one city where everything from politics, culture, economy, and science converged, except perhaps Philadelphia, where American chemistry got its start. But even Philadelphia wasn't on par with its European cousins. Eventually, solid relationships blossomed between the states as the country expanded and grew more interconnected. New societies and institutions helped feed an appetite for American science, colored with shades of Manifest Destiny. As Jefferson maintained, science and education would unite a nation where progress was the norm rather than the exception. Scientific innovation occurred as community truth-seekers built upon each other's work.

Most well-off Americans welcomed whatever was sensible, correct, and elegant. They generally eschewed the dubious, extravagant, and untrue.[35] Americans were keen to cultivate the practical element of the sciences. They recognized that science could have a great social impact. According to Alexis de Tocqueville: "Most people in such nations are quite intent on immediate material gratifications, and since they are always unhappy with the position they occupy and are always free to abandon it, they think only of ways to change or improve their fortunes.

To minds so disposed, any new method that shortens the road to wealth, any machine that saves labor, any instrument that reduces the costs of production, any discovery that facilitates or increases pleasures seems the most magnificent achievement of the human mind."[36]

This kind of philosophy influenced Benjamin Silliman to weave science into the national fabric. He knew that in the early 1800s American science remained a poor facsimile of the European model, and he hungered to change that. Silliman sought to understand Earth and all its imperfections.

"In ancient times, everything relating to natural bodies was included under Physics, and the term therefore comprised natural history, natural philosophy and chemistry," Benjamin Silliman said. "Since the more extended cultivation of natural science, and particularly since the time of Bacon and Newton, the external appearances of natural bodies have been included under natural history, their analysis and composition are assigned to chemistry. . . . The pupil learns, with pleasure and surprise, that the same power which retains Jupiter in its orbit, precipitates a falling drop; that a feather, a balloon and a ship of the line are floated by statical pressure."[37]

As America stood on the doorstep of the nineteenth century, people who might have once entered the military considered other professions as a way to affirm their patriotism. True, many material obstacles beset the United States, but the young nation transformed itself from an upstart grouping of thirteen colonies to an economic, social, and political force. The first seeds of scientific independence were sown.

Chapter Five

PRESIDENT JEFFERSON
AND NEW ENGLAND
ARGUE

S even years before the meteor crashed into the frozen farmland of Weston, Benjamin Silliman was busying himself with books and lectures at Yale College. It was 1800, and the first hotly contested presidential election between incumbent Federalist John Adams and his Democratic-Republican opponent, Thomas Jefferson, embroiled the country. George Washington's warnings about the corrosive nature of political parties had fast faded. Now two political parties, the Federalists and the Democrats, vied for power with unchecked enmity, polar opposites on every issue from commerce to religion. "It is now well understood that two political Sects have arisen within the U.S.," wrote Thomas Jefferson, "The one believing that the executive is the branch of our government which most needs support; the other that like the analogous branch in the English Government, it is already too strong for the republican parts of the constitution; and therefore in equivocal cases they incline to the legislative powers: the former of these are called federalists; sometimes aristocrats or monocrats."[1]

For the first time, rancorous partisanship seeped into American politics. Differences emerged concerning established religion's lack of tolerance for new scientific theories. The first mutterings of a profound dispute over the origins of existence were audible. Many in New England disagreed with Jefferson on the best way to govern. In the South more people supported the idea of strong states rights, and they didn't want the federal government to intrude. The idea of a federal government was new, and many were nervous about it; people trusted their own state governments, which had been colonies. The Virginia House of Burgesses had been running for two hundred years already. In contrast, many in the North favored a strong federal government. In many respects, it was the country's first culture war.[2]

The Connecticut Federalists, like Federalists nationwide, believed an Oval Office in which Thomas Jefferson sat behind the desk would destroy the American Republic because he and his administration would unravel the strong, central government and displace the authority of the churches. Founded in 1787, the Federalist Party advocated a strong federal government. In addition, the party pushed for a tariff system and a strong central bank. Although the party advocated mending the nation's relationship with Britain, for commercial reasons it practiced neutrality when it came to foreign affairs.

From the start it seemed as if most citizens of the state of Connecticut and Thomas Jefferson were destined to endure a fractious relationship. In 1801 Jefferson replaced Elizur Goodrich, a former congressman, with his own appointee, Samuel Bishop, to head the collectorship at the port of New Haven. To the chagrin of many New Haven merchants, Bishop supervised all revenues, fees, and taxes of the city. They regarded Bishop as Jefferson's pawn. More than eighty businessmen protested the appointment, saying it would financially ruin New Haven and Yale.[3]

Timothy Dwight, the de facto Federalist leader in the New Haven area, exaggerated the threat to Yale, but hyperbole served his purpose. The college president promoted a sense of urgency to further the Federalist Party agenda, which in turn fed the growing animosity between Jefferson and New England. As for Jefferson, he viewed New England as looking too favorably upon the British Empire and a stronghold of authoritarian churches. Aside from Dwight, the party's leaders included men such as John Adams, Alexander Hamilton, and Noah Webster, the latter a close friend of Benjamin Silliman.

Both pragmatic and religious, the Northeast fiercely opposed the man from Monticello. Regarding religion, Congregationalists dominated the

Federalist Party. They interpreted the Bible literally. Holidays revolved solely around prayer. Save for adorning windowsills with flickering candles, Congregationalists observed Christmas with minimal ornamentation. To do otherwise was considered a gaudy manifestation of papistry.

In 1800, the United States was essentially a Christian nation. States established government-supported churches. Until 1776 officeholders in the colonies professed their faith in "Jesus Christ, God as Savior." Even at the time of the Constitution's ratification, one had to swear a religious oath to serve. Not only were religious tests common, they were insulated from federal disestablishment. Over time, however, many Revolutionary-era leaders pursued Deism, which regarded Christ the man as simply a great teacher. They rejected supernatural events and refused the idea of miracles. Instead, Deists believed God gave humans the gift of reason. President Thomas Jefferson was a Deist and that offended Federalists. They didn't approve that Jefferson regarded the story of Adam, Eve, and the gleaming apple in the garden as merely a pretty tale. Federalists and Congregationalists believed the Almighty designed and implemented the world's wonders. Thomas Jefferson and his party did not.

Some of Jefferson's closest associates, with whom Silliman later clashed, suffered under the Adams administration (1797–1801). Thomas Cooper, a chemist, had been tried and prosecuted under Adams's Alien and Sedition Acts for publishing a series of editorials in the *Northumberland Gazette*. Cooper had laid into Adams and wrote of a litany of abuses by his administration. Men like Cooper had no shortage of ill will for New England. Their anger fed Thomas Jefferson's disdain for all things federalist. The rift grew.

During the bitter campaign of 1800, Timothy Dwight, nicknamed "The Pope," was Connecticut's leading clergyman. At forty-three, Dwight took satisfaction in his reputation as a well-known preacher, an excellent teacher, and a prominent poet. He graduated from Yale College in 1769 at the age of seventeen, and at nineteen he became a tutor at the school. In 1793 Timothy Dwight stood before the General Association of Connecticut, a missionary group that helped settlers in New York and Ohio. He advocated the need to defend the orthodox faith against Deists. As a principal of the rising evangelical faction of Congregationalism, Dwight had standing for both his skillful political leadership and his religious devotion. These traits served him well when he accepted the presidency of Yale in 1795.

Arriving at Yale, Dwight learned that less than 10 percent of the student body practiced religion outside the required daily prayers. He railed against this perceived campus-wide epidemic of religious and academic apathy. His efforts yielded results. By the time Benjamin Silliman graduated in 1802 and accepted his professorship, nearly one-third of the campus's 230 students professed their faith.

Gradually, under Dwight's rule, Yale fortified itself against the advance of Jeffersonian principles that New England Federalists abhorred. From his college president's bully pulpit, Dwight used the trinity of family, church, and school to rebuke the social disorder considered a hallmark of the Jeffersonian movement. On July 4, 1798, two years before the election, Dwight condemned Jefferson: "For what end shall we be connected with men of whom this is the character and the conduct? Is it that our churches may become temples of reason? . . . Is it that we may change our holy worship into a dance of Jacobin phrenzy [sic] and that we may behold a strumpet personalizing a Goddess on the altars of Jehovah?"[4]

Quite charismatic, Dwight invited controversy, but his community considered him one of its strongest pillars and most fervent patriots. "He is truly a great man, and it is very rare that so many excellent natural and acquired endowments are to be found in one person," wrote Silliman. "When I hear him speak, it makes me feel like a very insignificant being, and almost prompts me to despair; but I am reencouraged when I reflect he was once as ignorant as myself, and that learning is only to be acquired by long and assiduous application."[5]

On July 4, 1800, Timothy Dwight stood at his pulpit and inquired whether his congregants could "trust philosophers? Men who set truth at naught, who make justice a butt of mockery, who doubt the being and providence of God?"[6] This was one of many sermons Dwight delivered damning Thomas Jefferson as a Deist. He beseeched his flock to oppose the Virginian's candidacy. At a time when the clergy seemed to rule the countryside, Dwight criticized Jefferson over everything from his unorthodox views on science and culture to his beliefs on religion and commerce. Namely, the Yale president didn't approve of Jefferson's belief of the separation between church and state, his belief that it was necessary to study fossils in order to learn about the past, and his belief that people should practice whatever religion they wanted. In addition, Jefferson believed government shouldn't become the caretaker of the people, as the Federalists seemed to favor. "If we can prevent the government from wasting the labors of the people, under the pretense of taking care of them, they must become happy," Jefferson wrote.[7]

The hostility between New England and Jefferson intensified shortly before the 1800 election, particularly when the victory of the Virginia statesman seemed inevitable. The clergy and nearly every educated and respectable citizen of New England made no effort to conceal the loathing they harbored toward the incoming government. The barrage of accusations grew ever more forceful as Election Day drew near.[8] Religion inflamed political passions. Several early American newspapers, such as the New England *Palladium*, published anti-Jefferson pamphlets, which denounced his alleged atheism and his pro-French leanings. The Federalist press waged offensive, partisan assaults that would make modern Americans blush. They labeled Thomas Jefferson a pro-Revolutionary and the anti-Christ. They printed articles deriding the "Congo Harem" at his Virginia home, Monticello.

Ironically, it wasn't Thomas Jefferson's purported antics at Monticello that infuriated New England; they detested his alleged atheism, his foreign policy, and his desire for debate. "In every country where man is free to think and to speak, differences of opinion will arise from difference of perception, and the imperfection of reason; but these differences when permitted, as in this happy country, to purify themselves by free discussion, are but as passing clouds overspreading our land transiently and leaving our horizon more bright and serene," wrote Thomas Jefferson.[9]

The Federalists deemed him as someone who undermined their legally established, privileged, and tax-supported religion. In Jefferson, many citizens in New England perceived a man with zero regard for priests or revealed religion. To Timothy Dwight and his flock, Jefferson represented no less than the devil incarnate.

Intellectual elite among the Federalists twisted these popular senti- ments and used anti-intellectualism to smear the third president. They painted Jefferson as a deviant who pursued natural science and philos- ophy, something only a heretic would dare to do. Each day for two months before the election the lead Federalist newspaper, the *Gazette of the United States*, posed the question: "The Grand Question Stated! Do we vote for Adams, and a preservation of religion in America or do you vote for Jefferson and no God?"

Newspapers began to play an important role in the public life of the new nation. At the turn of the century, the United States could boast the existence of more newspapers with a combined circulation larger than any other country in the world. Alexis de Tocqueville remarked that the American press circulated politics through the nation, its power second only to that of the people.

A former schoolteacher named John Fenno breathed life into the *Gazette* in 1789. Although Fenno, a Federalist Party activist, attended Holbrook's Writing School in Boston, he never really mastered the craft. He let others brandish the pen and, until he died from yellow fever in 1798, he was a publisher devoted to printing the glories of federalism. In the years before the American Revolution, activists used newspapers such as the *Gazette* as platforms to advocate separation from colonial rule. As the infant nation grew, political activists on both sides made full use of the power of the press, only this time to further their own political ambitions. The *Gazette of the United States* faced off against Peter Fre- neau's *National Gazette*, as did Noah Webster's *Aurora* and William Cobbett's *Porcupine Gazette*. None of these papers let truth obstruct their streams of vitriol, and all of them unflinchingly engaged in verbal fisticuffs.

Each time a new Republican publication rolled off the press, Feder- alist editors read every article as a personal challenge. Likewise, the Republican editors interpreted each Federalist article as a personal affront. After they finished reading the opposition, they returned to their own press, determined to print the last, if not the most level-headed, word.

Every paper took a side during an election. When Thomas Jefferson publicly professed his idea that religion should be private, and that people should be free to practice a faith of their choosing, New England boiled. Those who challenged Jefferson decreed that if he occupied the newly constructed White House, he would incite God to visit revenge upon the new Republic: "Can serious and reflecting men look about them and doubt that if Jefferson is elected, and the Jacobins get into authority, that those morals which protect our lives from the knife of the assassin— which guard the chastity of our wives and daughters from seduction and violence—defend our property from plunder and devastation, and shield our religion from contempt and profanation, will not be trampled upon and exploded."[10]

Then Benjamin Silliman's close friend Noah Webster entered the political ruckus. Webster, already the author of several textbooks and political pieces, had founded the newspaper *American Aurora* on December 9, 1793. He condemned Jefferson's religious experimentation: "Never let us exchange our civil and religious institutions for the wild theories of crazy projectors; or the sober, industrious moral habits of our country for experiments in atheism and lawless democracy," Webster wrote. "Experience is a safe pilot; but experiment is a dangerous ocean, full of rocks and shoals."[11]

A season of tongue lashing showered the nation. Abusive attacks reigned, and newspapers eagerly published the bitterest, most vituperative slurs. A Federalist contributor to the newspaper *Hudson Balance* undertook to prove two propositions: "First that Mr. Jefferson was an infidel. And secondly, that he would be pleased with a subversion of Christianity in this country."[12] In the *Connecticut Courant*, the editors unleashed this tirade: "Look at your houses, your parents, your wives, and your children. Are you prepared to see your dwellings in flames, hoary hairs bathed in blood, female chastity violated, or children writhing on the pike and halberd? . . . Look at every leading Jacobin as at a ravening wolf, preparing to enter your peaceful fold, and glut his deadly appetite on the vitals of your country . . . GREAT GOD OF COMPASSION AND JUSTICE, SHIELD MY COUNTRY FROM DESTRUCTION."[13]

Connecticut regarded Jefferson as Jeroboam, the king who led Israel into sin because he appeared to doubt the authority of revelation. The cadre of New England clerics twisted Jefferson's ideas of religious tolerance: "Let my neighbor once persuade himself that there is no God and he will soon pick my pocket, break not only my *leg* but my *neck*. If there be no God, there is no law, no future account; government then is the ordinance of man only, and we cannot be subject for conscience sake."[14]

Aside from attacking Thomas Jefferson's religious views, New England Federalists assailed his scientific pursuits. They labeled both the president and his supporters as "philosophical." The word, actually code, appeared whenever a newspaper or broadside planned a particularly stinging diatribe against Jefferson or his party. The word became synonymous with everything that Thomas Jefferson championed, from the French Revolution to the study of fossils, which seemed frivolous.

Nevertheless, Jefferson won after the House of Representatives decided the election. There was a tie in the Electoral College between Jefferson and Aaron Burr. As a result, Congress decided the outcome. After Election Day many towns and cities across the country celebrated the successful candidate with parades, church services, feasting, turkey shoots, and the serving of an Election Day cake, a spiced fruitcake. Needless to say, Connecticut and much of New England refused a slice.

Federalist papers such as the *Palladium* editorialized: "Should the infidel Jefferson be elected to the presidency, the seal of death is that moment set on our holy religion, our churches will be prostrated and some infamous prostitute under the title of the goddess of reason, will preside in the sanctuaries now devoted to the worship of the most high."[15] In that vein, a thick black line bordered the front page of Boston's *Columbian Centinel* like a funereal ribbon.

In contrast, Webster's *Aurora* in Philadelphia published this editorial on March 9, 1801: "The sun of Federalism sets this day—would that its deleterious influence were to die with it. But sufficient for the day is the evil thereof."[16] The *National Intelligencer* also celebrated: "As far as accounts have been received from the various parts of the union, the election of Mr. Jefferson to the presidency has produced the liveliest feelings of joy. In Baltimore, Philadelphia, and New York, the bells have been rung, the artillery been fired, and convivial entertainments given."[17]

As Thomas Jefferson ascended to the highest office in the land, Connecticut steeled itself against a man they believed would marginalize their core beliefs. Economics also fed the struggle between Jefferson and New England. The merchant class of New England opposed Thomas Jefferson's election. They were convinced that his policies would cramp their business. This rancor led President Jefferson to seek revenge. After his inauguration he purged Federalists from office. By the time Thomas Jefferson took the presidential oath of office on March 4, 1801, the seeds for a terrible mistrust had been planted.

Although the 1800 elections swept the Federalists out of nearly every office, Connecticut Federalists retained all of their seats in Congress. Governor John Trumbull of Connecticut survived the election and helped the party remain vocal. Chief among the party's worries was its impending loss of power and prestige. The federal capital was being relocated to the drained swamplands of the Potomac, a move that alienated citizens of New England. They felt that the nation's power base was shifting to the "infidel, anti-commercial, anti-New England Southerners."[18]

Much of Connecticut's insecurity could be traced back to the fact that with Jefferson's election, the Northeast would wield much less control than in the previous generation. Until the War for Independence, power sat on a gilded throne across the Atlantic. How many people lived in a particular colony didn't matter then. But now it did. As lands to the west opened and as the South began to rise, the balance of power shifted. For example, Northerners feared that Southerners would settle the vast territories that lay to the west, thereby taking one more step toward permanently consigning New England to minority status in the Union.

New Haven and other towns along the New England coast also dwelled in conservatism because of the economy. The New England ruling class feared that freethinking and free commerce would diminish their economic power. This insecurity fed Connecticut's Federalism. John Adams, the second president, fully empathized. He understood the Nutmeg State's social and economic concerns. "The state of Connecticut has always been governed by an aristocracy, more decisively the empire of Great Britain. Half a dozen, or, at most, a dozen families, have controlled that country when a colony, as well as since it has been a state," said John Adams.[19]

Ironically, the ability of Connecticut Federalists to sustain power in their state did not make them feel more secure. Rather, they felt they were a fortress of rational liberty, severed from the rest of the nation and beset by an alien Jeffersonian culture. In fact, the feeling of living under siege grew so strong that many Northern Federalists toyed with secession. They justified this flirtation with a divorce from the Union after convincing themselves they were merely breaking from Virginian hegemony. Noah Webster, a graduate of Yale, led this separatist movement together with Jeddidiah Morse, the father of Samuel Morse and a clergyman, as well as Governor Jonathan Trumbull and Yale president Timothy Dwight. Although Benjamin Silliman was Webster's close friend and Dwight's student, there are no records of his views on secession.

Midway through Jefferson's first term, Deism's popularity waned. An evangelical tide washed over the country. As this religious revival swept Yale College in 1802, Benjamin Silliman reflected on the changing climate. He wrote in a letter to his mother: "Yale College is like a temple—prayer and praise seem to be the delight of the greater part of the students while those who are still unfeeling are awed into respectful silence."[20]

Even so, during this time Timothy Dwight despaired over the moral tone of both the college and the nation. He believed orthodox religion was declining and that Deism and skepticism had grown too powerful. A true Connecticut aristocrat, Dwight failed to appreciate how much frontier settlements embraced the spirit of freedom and equality. He didn't understand that Congregationalism contradicted classlessness. He didn't see that many Americans flirted with atheism and had adopted a firm indifference to religion. Moreover, Dwight bemoaned the influence that the French Revolution seemed to possess in parts of America.

Thomas Jefferson believed religion was a personal matter. He thought European churches exploited irrational superstitions to oppress people and secure the thrones of despots. He ached for empirical science. Even more irksome to New England, Jefferson rarely attended church services and he actually worked and entertained on the Sabbath.

Discontent plagued Timothy Dwight as he walked the campus green of Yale. He regretted the country's decline of piety; even his flock increasingly ignored the strict moral code he espoused. "The principal amusements of the inhabitants are visiting, dancing, music, conversation, walking, riding, sailing, shooting [at a mark] draughts, chess, and unhappily in some of the larger towns, cards and dramatic exhibitions," Dwight wrote.[21] He likened the reading of newspapers to drinking, swearing, and gambling. If one had to read, then only the Bible should be read. He frowned on theater but approved of corn husking, folk songs, psalms, ballads, and the occasional ride in a sled.

Timothy Dwight constantly warned Yale students to shun Deistic philosophy; he and Connecticut Federalists wanted to preserve and conserve the state's traditional institutions. Dwight believed that Connecticut should remain homogeneous: "In this country you all sprang from one flock, speak one language, have one system of manners, profess one religion, and wear one character. Your laws, your institutions, your interests are one. No mixture weakens, no stranger divides you."[22]

The Congregationalists, and the Federalists, grew increasingly hysterical and vituperative. They abhorred Thomas Paine's attempt to seduce America with his pamphlets of the 1790s titled *Age of Reason* and *Rights of Man*. They opposed Jefferson's vocal admiration of these publications, which called for people to question revealed religion and seek greater political freedoms. Federalists convinced themselves the third president and Paine actually served France. In the late 1780s, the French public lauded these ideas. During the French Revolution the National Assembly usurped church property and abolished the observation of the Christian Sabbath. These actions horrified New England Federalists, who believed that Jeffersonian Republican-Democrats aligned themselves with the radical factions of Revolutionary France. The Federalists believed Enlightenment thought detonated the violent uprising against King Louis XVI. New England pointed to the storming of the Bastille and said it proved that the masses were incapable of self-government. The Federalists insisted that a strong executive with concentrated power was the only rational choice. A conservative Puritanism thundered from Connecticut's pulpits, and control of Yale enhanced Congregationalism's power.

Jefferson's detractors claimed that he rejected religion. Not so. Although as a young man Jefferson rejected his parents' Anglican faith, he simply believed that religion was the master key to virtue and morality; its values necessary to build a successful republic. He actually affirmed religion, albeit for pragmatic and utilitarian motives.

Like many, he viewed science through the prism of patriotism as well as through the lens of religion. Deists believed in "reason" and "nature." Thomas Jefferson also ardently believed in the separation of church and state. Theology had no business in politics. Furthermore, he considered the Bible to be little more than a history of humanity rather than the revealed word of God.[23] Like other men and women touched by the Age of Enlightenment, Deists believed God couldn't intervene to counter the natural laws. Jefferson discarded the notion of an arbitrary, jealous, and vindictive deity. He was more interested in improving life in the present than in turning an eye toward reward in an afterlife.

Yet Jefferson refused to answer charges of atheism in the press: "I determined never to put a sentence into any newspaper. I have religiously adhered to the resolution through my life, and have great reason to be contented with it." He wrote to his friend Samuel Smith that if he answered every allegation, he would have no time for anything else: "for while I should be answering one, twenty new ones would be invented."[24]

Some New Englanders labeled Jefferson's interest in science both un-Christian and insulting. New England's clergy dreamt of cultivating a land reflective of Puritan philosophy, but it overstated the situation to contend that they shunned all rational thought. Many clerical leaders, like Timothy Dwight, were curious and open to scientific theories. Though a devoted Congregationalist, Dwight never stopped pursuing science. His very appointment of Benjamin Silliman demonstrated his understanding of science's import. He discussed climate and geology in his book *Travels in New England and New York*. Still, he believed science should further disclose the ways and wonders of God as revealed in the Bible because he believed everything answered to God.

Now a professor at Yale, Benjamin Silliman spent more time shaping what he stood for, not what he stood against. The young professor didn't share Federalist skepticism of Jeffersonian science. He, too, believed in invention, anthropology, and rational explanations for seemingly myste-rious phenomena. His definition of the proper mission of scientific investigation differed greatly from many Federalist-Congregationalists who considered Jefferson's ideas dangerous. Indeed, many strict Congregationalists were in a quandary; accepting science meant embracing ambiguity and ignoring the credo of divine creation. It wasn't that New Englanders opposed exploration. Most clergy accepted scientific teaching and research; they just viewed the discoveries as evidence of their Lord's might.[25]

For his part, Benjamin Silliman showed a remarkable capacity to balance his religion with his pursuit of science. He helped close the fissure between the revolutionary era and the Age of Enlightenment, while at the same time crafting an intellectual bridge between religion and science. He moved easily between Deists and Congregationalists, Federalists and Democrats.

As Yale's professor of chemistry and natural history, Benjamin Silliman wholeheartedly accepted the Bible as a revelation of truth and the source of ultimate authority. He believed the Bible contained a moral code. He believed that its teaching agreed with geological time, beginning with the appearance of the seas and ending with the arrival of people. Silliman acknowledged the sensitivity of his home region, where religion sometimes overpowered science. He carefully negotiated and accommodated views the Congregationalist Church didn't wholly approve. Silliman sincerely respected his religious and secular teachers. He always showed them courtesy and restraint whenever he published anything about geology.

Nonetheless, Silliman firmly held that none of the natural sciences, geology, chemistry, or paleontology contradicted the Old Testament. He interpreted the Bible as a book intended to teach ancient people about religion. In his eyes, the tome had nothing to do with teaching modern people about science.

"Religion, the parent of moral science, is derived, in part from nature justly viewed; but, its principal, and only infallible source, is the direct revelation which it has pleased the Almighty to make, through the gospel of reconciliation," Silliman said. "It is neither superfluous, nor impertinent, to mention it on the present occasion. It is the head of all sciences; it is the only revealed one, and it is necessary, to give a proper use and direction to all the others."[26]

Benjamin Silliman stood for possibility. American science was young, and some religious elements threatened its growth. Men like Isaac Bronson, Nathan Wheeler, Benjamin Franklin, and Thomas Jefferson proved there didn't need to be a fissure between science and scripture. Patriotism was consistent with piety and practicality. These men believed one's duty was to acquire a command of nature, science, and practical invention. They believed government should aid, not thwart, discovery.

Silliman helped America advance by enabling its indigenous enterprises to profit from the country's many natural resources. Patriotism, utilitarianism, and a love of science and scientific reputation sparked the imagination of the first generation of American scientists.

"Among the privileges which the Deity has conferred on man, one of the greatest is the capacity and the disposition to acquire knowledge. Our species is thus distinguished from the various orders of inferior animals," Silliman said. "But, to the advancement of the human mind, there are no limits: The world is all before us, where to choose, and Providence our guide."[27]

Chapter Six

WHEREBY SILLIMAN AND KINGSLEY CONVINCE THE PUBLIC

O nce upon a time American essayist Washington Irving said this about the inhabitants of New England: "The people at large show a keenness, a cleverness and a profundity of wisdom that savors strongly of witchcraft—and it has been remarked that whenever any stones fall from the moon, the greater part of them is sure to tumble into New England."[1] Except, the stones didn't fall from the moon. Professor Benjamin Silliman proved that.

After completing his fieldwork in Weston, Silliman returned to Yale. As usual, when initially facing a significant task, he turned to his mother. Silliman revered her as a "heroic woman" whose courage, devotion, and piety had laid the foundation for his own achievements.[2] He sought her advice many times during the next several weeks. "My dear mother," Silliman wrote from Yale College on January 10, 1808, "I have been acidulously [*sic*] employed, night and day, for a week past in analyzing the stones . . . to ascertain the exact proportions as well as nature of the ingredients and have succeeded to my own satisfaction. I shall publish the account both here and in Europe."[3] If he published first, Silliman could convince Europe that the student—the United States, in this case—had surpassed the master. Then Yale president Timothy Dwight would know that he had chosen well five years earlier when he asked Silliman to become the college's first professor of chemistry.

Silliman cloistered himself inside his laboratory. Relevant literature from chemists, geologists, and others on the subject of meteorites crammed his workspace. He frequently consulted these publications to both refute and confirm his developing hypotheses. He used certain accepted analytical methods to formulate them. He knew many European and American scientists persisted in believing that lunar volcanoes disgorged meteorites. He knew other scientists maintained the view that meteorites originated in the atmosphere.[4] Silliman staked his reputation on the outcome of his work. He and his colleague James Kingsley knew they needed to act swiftly to prove their point, namely, that meteorites came from outer space.

Silliman and Kingsley pursued their work like any scientific endeavor; they built upon previous successes and failures. Thus, Silliman referred to Immanuel Kant's 1755 work, which stipulated that everything in the solar system, from the planets to the comets, came from materials that once filled the entire universe. Although Kant was a philosopher, his work provided Silliman with a starting point for his thesis. Silliman also elaborated upon the work of American scientists such as Andrew Ellicott, an astronomer, surveyor, and instrument maker. Ellicott, the son of a Quaker clock maker, had done many early calculations regarding the rising and setting of heavenly bodies. These calculations helped Silliman when he mapped the fall and when he tried to fix the location and speed of the meteorite as it traveled through the sky.

Silliman read the work of John Winthrop of the Massachusetts Bay Colony. Winthrop corresponded with the Royal Society in London about the appearance of comets in the late 1700s. David Rittenhouse, who succeeded Benjamin Franklin as president of the American Philosophical Society in 1791, correctly speculated that meteorites were extraterrestrial bodies, though he didn't know where they came from or how they fell to Earth.[5] Thomas Clap of Yale authored *Conjectures upon the Nature and Motion of Meteors, Which Are Above the Atmosphere*. The discourse was published after Clap's death in 1781. In this work he posited the view that meteors were terrestrial comets. Clap claimed that nothing else could pitch bodies hundreds of miles above Earth's surface while simultaneously giving them a projectile velocity of twenty times that of a cannonball. He said they measured about half a mile in diameter and were encased by a solid exterior. Although Clap erroneously believed meteorites orbited Earth, he correctly surmised that friction with Earth's atmosphere caused the loud noises most witnesses reported. Time and again Benjamin Silliman and James Kingsley probed Clap's theory while laboring over their own paper. "I have had on occasion many times to detail and illustrate President Clap's theory, and it has generally been considered as better than any other; the lunar philosophers are humorously called lunatics," Silliman remarked to Kingsley.[6]

Silliman and his Yale colleague Professor Kingsley drew on the work of Ernst Chladni. Born in Wittenberg, Germany, in 1756, Chladni's interests included mathematics, physics, music, and natural history. He became fascinated by meteors after meeting Georg Christoph Lichtenberg, one of Europe's preeminent natural philosophers and physicists. Their meeting spurred Chladni to compile historical reports about twenty-four fireballs and eighteen falls. In 1794, Chladni published a book titled *On the Origin of the Mass of Iron Discovered by Pallas and Others Similar to It and on Some Natural Phenomena Related to Them.* The slim volume suggested the stones were indeed celestial bodies born in space. However, Chladni never actually observed a fall or spoke with a witness. He based his entire work on second- and third-hand accounts. Moreover, his supposition fully contradicted the prevailing belief that absolutely nothing could fall from the sky.[7]

Enter Professor Benjamin Silliman, the first scientist to interview eyewitnesses and personally analyze the stones that composed a meteorite. In his laboratory, Silliman executed many experiments to identify the meteorite's composition. The largest unbroken chunk weighed about twenty-six pounds. Because he read so many historical accounts, Silliman knew to search for the element nickel. A chemical test for nickel is definitive for meteorites about 99 percent of the time. All iron meteorites, and most all stony meteorites, contain some percentage of nickel. In addition he found silex, iron oxide, magnesium, oxide of nickel, and sulphur. Aside from magnesium, silicon is the principal ingredient in meteorites. Silliman compared the stone's elements with terrestrial rocks. The principal difference is that terrestrial rocks aren't attracted to magnets.

Silliman cracked open the stones he harvested during those frosty December days in Weston. Much to his delight, he discovered they were studded with celestial gems. The nearly spherical beads are called *chon-*

drules. They take their name from the Greek word *chondros*, meaning "grain." Composed of silicon, the granules range in size from those as small as a pinhead to others the size of peas. "Often the grains are so large and distinct that the exterior of the stones, usually of an ash grey, presents the aspect of certain breccias,"[8] wrote Silliman in his notes on the meteorite.[9] The chondrules excited him. He knew earthly rocks don't contain these globular pieces. He looked upon these tiny spherical inclusions as proof that the Weston meteor hailed from the great beyond.

For more than a century, scientists argued passionately over how, and from what, these mysterious little jewel-like spheres are formed. Geologists and meteoricists now agree that chondrules contain the seeds of the solar system. With so many chondrules, the Weston meteorite earned its place as an exceptionally important specimen that held clues to Earth's origin.[10] Later scientists would determine that the Weston meteorite likely began its journey to Earth some thirty million years ago after two asteroids collided. The collision produced thousands of rocky bodies, further crowding the asteroid belt between Mars and Jupiter. Generations after Silliman, scientists would learn that over time, gravitational forces from different planets tugged the object to and fro, thereby altering its orbit. It crashed to Earth on December 14, 1807.

The Weston meteorite is classified as an H4 chondrite. The H stands for "high iron." Meteorites like that found at Weston contain about 25 to 31 percent total iron and are easily attracted to magnets. It is neither the most unique nor the most ordinary. Chondrites rank among the most primitive samples available to scientists, allowing them to study the solar system's earliest history. In the twentieth century, scientists realized similar materials formed Earth. So when Benjamin Silliman studied the Weston meteorite, he studied Earth's earliest history.[11]

As Silliman methodically chipped away the specimen's layers, he neatly recorded his observations. He diligently reviewed the testimony of Judge Wheeler and Elijah Seeley. He noted Elihu Staples's observations about the meteorite's flight path. Every day Benjamin Silliman spent sequestered in his lab scrutinizing the rock, the premise that lunar volcanoes hurled meteoric stones weakened. Meteorites weigh at least twice as much as terrestrial rocks. Consequently, a greater proportional force must throw them, wrote Silliman. The professor found the lunar volcanic theory weak because of the presence of iron. He said a volcano could never generate enough velocity to hurl a rock beyond Earth's circumference and through its atmosphere. Either the object would revolve around Earth for eternity or fly into unlimited space.

In January of 1808, when Benjamin Silliman again wrote his mother, his stress was apparent. "Copying the account of the analysis to be published in New York, Philadelphia, London, and Paris detains me and now fills my time completely. I hope you will not be anxious my dear mother. I am perfectly well and my brain is regular notwithstanding."[12]

The Silliman-Kingsley work generated incredible interest. Local papers published accounts of the incident under the heading "Terrestrial Comet." Silliman penned these articles, which were early drafts of the more scholarly paper he was preparing. The *Connecticut Herald* published the first article by Silliman about the Weston Fall on December 29.[13] The article put Silliman and Kingsley on record. It also summarized the eyewitness accounts and offered interested readers the chance to read the preliminary results of Silliman's chemical analysis. His chemical report snared the public imagination; it was the first step toward thoughtful popularization of science.

The January 1808 issue of *Churchman's Magazine*, whose subtitle identified it as a "Treasury of Divine and Useful Knowledge," reprinted the Silliman-Kingsley article. It included an introduction intended to preempt alternative accounts. The authors acknowledged that erroneous accounts were finding their way into print. To straighten the record, the pair explained how they visited and examined every spot where stones fell. They told readers how they collected specimens and "conversed with all the principal original witnesses; spent several days in the investigation, and were, at the time, the only persons who had explored the *whole* ground."[14]

Moreover, the publication's editor, the Reverend Tillotson Bronson, explained the meteorite's ultimate significance:

> As Christians we believe that all events in the natural world are subject to the control of Almighty Power. . . . But what is this compared with the events of the last great day when planets and suns shall be hurled from their spheres and the earth not only shaken, but from her centre torn to be involved in flames and dissolved into smoke and vapors? . . . Think of this, O Christian! and consider what manner of person thou oughtest to be, *in all holy conversation and godliness.*[15]

Furthermore the two professors argued for a specific hypothesis about the meteorite's origin.

> The chemical analysis also proves that their composition is the same; and it is well known to mineralogists and chemists, that no such stones have been found among the productions of this globe. These considerations must, in connection with the testimony, place the credibility of the facts said to have recently occurred in Weston, beyond all controversy. To account for events so singular, theories not less extraordinary have been invented. It is scarcely necessary to mention that theory which supposes them to be common masses of stone fused by lightning, or that which derives them from terrestrial volcanoes; both these hypotheses are now abandoned. Their atmospheric formation, from gaseous ingredients, is a *crude* unphilosophical conception, inconsistent with known chemical facts, and physically impossible.—Even the favorite notion of their lunar-volcanic origin, seems not to be reconcilable with the magnitude of these bodies, and is strongly opposed by a number of other facts.[16]

Benjamin Silliman's description of the fall and his chemical analysis of the stone meteorite, the first performed in this country, received an incredible amount of attention. His paper received the most comment, both support and criticism, of any scientific paper of its time. The chemical analysis alone remains historically valuable because it is the earliest-known research of this type in the United States. Indeed, the media attention advanced the growing celebrity of the Yale professor.

Despite the mounting evidence about meteorites' origins, several detractors held firm. Many a naysayer came from Silliman's own circle of colleagues. These critics included Dr. John Brickell of Charleston, South Carolina. Brickell said if meteorites came from outer space, their weight would slowly increase the mass of Earth. This would cause Earth to revolve closer to the sun. And if that happened, then other bodies in the solar system, including the moon, would change their orbit.[17]

Brickell had company: Silliman's former teacher Professor James Woodhouse of the University of Pennsylvania craved his share of publicity. David Judson, Silliman's Fairfield friend, sent word that Woodhouse desperately tried to secure fragments of the meteor. Woodhouse boasted to Judson that "if he should not recover them, no truly accurate account of the meteor could possibly appear."[18] Woodhouse advertised himself as the sole individual capable of analyzing the rock. Judson was foremost a loyal Silliman family friend, so news of Woodhouse's letter reached Yale.

The letter angered Benjamin Silliman. He decided to travel to Philadelphia to safeguard his scientific proprietorship of the stone. He also wanted to confront his former teacher. Both Woodhouse and Silliman exchanged letters before the latter departed. Woodhouse tried to mollify Silliman. He claimed he had no inkling that Silliman was working on the stone. Whatever Silliman said must have achieved his desired effect, because Woodhouse later apologized.

> Your statement is more accurate than any other which has as yet been given to the world . . . when I wrote to Mr. Judson that if these stones did not fall into my hands, no knowledge of their component parts would be given to the world I was ignorant that you had paid any attention to this subject. I have examined the parts of the stone attracted by the magnet and the unattracted portion. . . . [P]arts appear to consist of iron, sulfur and nickel. . . . When you arrive in Philadelphia, I will have more conversations with you on this subject, Yours sincerely.[19]

It was, however, a qualified apology. Along with the letter Woodhouse had enclosed his own analysis. The Pennsylvania professor couldn't refrain from trying to attach his name to the first official report.

Once again Benjamin Silliman found himself on the rough road to Philadelphia. He hired a coach from New Haven and spent the night in Fairfield, where he conferred with David Judson. He arrived in New York the next morning. In Manhattan, Silliman enjoyed the hospitality of fellow Yale alumnus Benjamin Perkins, a collector of gems and minerals. As Silliman traveled to New Jersey, the temperatures remained glacial, matching his cold fury toward his former professor turned foe.

"We arrived on Wednesday morning, after riding all night through New Jersey. . . . [Woodhouse] received me politely, but made no allusion to the offensive part of his letter," Silliman wrote to Kingsley, who stayed behind in New Haven. "The meteor is immediately brought forward in every circle where I go . . . it was so at Woodhouse's. I dined with him yesterday and met a large party of swans."[20]

No record survives as to the offensive part of the letter, but conversation focused on the meteor wherever Silliman went, even during the frosty dinner at Woodhouse's home. Several established scientists sat around the dinner table, including Adam Seybert, who befriended Silliman on his first trip to Philadelphia in 1803. Among the other guests was Dr. Thomas Cooper, a close friend of President Thomas Jefferson.

As they dined, Seybert warned Silliman to be ever mindful of Professor James Woodhouse. "[His] reputation is *up*, both here and [in] New York, for unfair dealing and in matters affecting scientific reputations. . . . I am convinced, from what [Seybert] has told me, that [Woodhouse's] own analysis was altogether loose and not to be depended on, nor am I at all afraid of any publication of his. Seybert advised me not to trust him; said he would play me some trick, for instance, purloin and publish it as his own; and averred that he did not know how to analyze a stone, and that he had not a single sure test or agent of any kind to do it with."[21]

Thomas Cooper's presence, let alone his running commentary on everything from religion to science, irritated Silliman. Finally, Silliman had enough.

"With an air of ridicule and self-importance, [Cooper] began questioning me, and intimated incredulity on several chemical and astronomical points; but I met him with a decided[ness] and severity which I would not often indulge in society, and the Doctor being really as ignorant as he was vain and impertinent. I found no difficulty laying him on his back," Silliman wrote to James Kingsley.[22]

His trip to the City of Brotherly Love paid off, nevertheless. Silliman's growing ability to translate the language of academia into the parlance of laypeople helped convince the public that the Weston meteorite was not of this world. People questioned him about the amazing event at nearly every turn. Silliman's work whetted the nation's appetite for science. "You don't know how keen the world this way are [*sic*] for meteors," Silliman said.[23]

Publications, such as pamphlets and tracts, had begun to popularize science in 1543, when Copernicus published *De Revolutionibus Orbium Coelestium* (On the Revolutions of the Heavenly Spheres). In 1633 the Catholic Church accused Galileo of heresy, not so much because of his argument but because he popularized his heliocentric model of the solar system in *Sidereus Nuncius* (Starry Messengers). A 1690 medical story about plagues and agues became one of America's first science stories. Now newspapers and magazines spread the word of the Weston Fall; anyone who read could enjoy science. Literacy rates rose steadily at the beginning of the nineteenth century. In some places in New England, the rates exceeded 90 percent.

Several days passed after the Woodhouse incident, and Benjamin Silliman once again ensconced himself in his lair. He turned to the manuscript, working diligently to ensure a pristine report. The *Connecticut Herald* already ran preliminary findings, but the scientific community still hadn't officially reviewed the report. So while Silliman cared about public reaction, he cared more deeply about his colleagues' recognition that he was right. Any error or mistake in either his experiments or his paper would have disastrous effects on his career and his reputation.

Dr. Isaac Bronson suggested Silliman and Kingsley collaborate on a popular book. Bronson thought the two scientists should collect "all well-authenticated instances of similar events, arrange and illustrate them, relate our own case with the analysis, and the result of the analytical examination of the rest . . . in short, make a book." He proposed they sell the work for one dollar a copy. Silliman warmed to the idea and told his colleague Kingsley they "could write into reputation and bread."[24]

Silliman decided that Kingsley should handle the section on meteorite origins. He, on the other hand, would deal with all scientific matters. "You must collect all the historical evidence. I will do everything connected with the mineralogy and chemistry, and together, and with the occasional aid of Brother Day [Yale colleague Jeremiah Day], we will state and refute the prevalent theories and magnify our own and make it honorable."[25]

Again Silliman shared his private musings with his mother. "The general opinion here is however that . . . there can be no question that my book will sell. . . . [A]nd if it has merit it will be a permanent source of profit. The crisis of the times he thinks makes no difference whatever. I am greatly encouraged by his opinions and shall commence the work immediately on my return."[26] Silliman thought the economic downturn as a result of Jefferson's Embargo Act wouldn't prevent people from buying his book.

After intense deliberation, Kingsley and Silliman opted for academic credibility over financial gain. Silliman asked Jeremiah Day to assist in writing the report. Day, a professor of mathematics and natural history at Yale, acknowledged the controversy surrounding the origin and composition of meteors. He respectfully mentioned those who believed meteorites were volcanic in origin. To further buttress this portion of Silliman's report, Day contacted his brother Thomas, who lived in Hartford. Several weeks later, Thomas wrote back telling his brother about several reliable eyewitnesses.[27]

Silliman submitted the report to Franklin's American Philosophical Society in Philadelphia, the offshoot of his earlier club, Juno. In the mid-eighteenth century, the American Philosophical Society attracted sparkling ideas and stellar minds. Its early members included George Washington, John Adams, Thomas Jefferson, Alexander Hamilton, Thomas Paine, and David Rittenhouse. Membership dwindled until 1767, when the philosophical society joined the American Society for Promoting Useful Knowledge.

The palpable interest from the society put Silliman in a bit of a quandary. He couldn't decide whether to present his results to a European audience before allowing the Philadelphia group to publish the findings in their journal, *Transactions*. In his opinion, the Old World seemed more receptive to accepting the theory that meteors came from outer space. Even his friend Dr. Robert Hare urged Silliman to send his articles to British journals. Hare contended that publishing abroad would boost Silliman's reputation.

But Silliman felt duty-bound to the United States and the advancement of its science. Still, he and Hare harbored some derision for the American Philosophical Society. Both Silliman and Hare said the society hadn't satisfied the emerging scientific class. It often overlooked those who lacked either familial or fiscal connections to the city's patrician class. Silliman disapproved of associates who purchased their membership. He believed the American Philosophical Society needed to be more discerning with respect to its admissions.

Benjamin Silliman debated where in the United States to first publish: Boston or Philadelphia. There were differences. The Boston-based American Academy of Arts and Sciences emphasized mathematics and astronomy over natural history. In contrast, the Philadelphia-based American Philosophical Society emphasized botany, geology, and paleontology. Silliman had an affinity for Philadelphia, where he had experienced a scientific awakening. Politics also played into the decision. The American Philosophical Society was considered Democratic, and Boston's academy was viewed as a Federalist stronghold. Federalists tended to distrust the society, no matter what it published, particularly anything about natural history.

Silliman ultimately agreed to publish with the American Philosophical Society. On March 4, 1808, he read the meteorite account before the society's committee. Attending members included Professors James Woodhouse, Robert Hare, and Joseph Cloud. If Woodhouse felt bitter, he hid his personal feelings. The paper was a success.

Silliman's endeavor offers rare insight into the state of American science as it stood on the cusp of a new international identity. The Weston Fall coincided with America's growth in politics, economics, science, and the arts. Trying to forge a new relationship with Europe and the rest of the world, the United States wanted to be seen as both an independent nation and a strong nation.

Europe spurned American science. The Old World long considered American science as hot air and gibberish. Then along came a blazing meteorite, Benjamin Silliman, and his detailed report. After Frederich

Hall, an American living abroad, published a copy of Silliman's report, Europe finally cast a new eye toward the young nation.

"Dear Sir, I lately received an American newspaper containing your relation of the phenomenon, which happened at Weston. I read it with much interest and pleasure," Hall wrote Silliman.

> I handed it to Count Rumford. He gave it to Mr. Pictet, one of the inspectors of institutions, who translated and read it, on Monday last, before the first class of the National Institute. I have occasionally attended the meetings of this class for more than six months, and I assure you, I have never heard anything read which ingaged [sic] more attention, or which appeared to be better received. . . . Most of the American productions, which reach Paris, even those on the sciences, are written in a style of high bombast, in language indefinite, and often unintelligible. From these the French form an unfavorable idea of the state of learning in our country. They ridicule, and very justly, Dr. Knott's description of the ellipse of 1606. Your account is minute and simple and therefore more credible and more suitable for a subject of this nature. When Mr. Pictet finished reading it, I heard from *la foule* [the crowd] of this literati, the expressions, *C'est une bonne relation. Elle est bien faite.* [It's well told. It is well written.] I felt a real pleasure in witnessing this approbation, for I considered it honorable to our country, (which is not much burdened with foreign literary honors) as well as to yourself and Mr. Kingsley.[28]

When the French journal *Annales de Chimie* printed "Note sur la chute d'un aerolithe a Weston" (Notes on the Meteorite Fall in Weston), it elevated Silliman. For the first time, an American scientist shared the intellectual stage with his European counterparts.

Once Silliman published, other scientists eagerly put forward their own counter-theories. The Philadelphia Medical Museum printed Professor James Woodhouse's article, which was considerably briefer than the Silliman-Kingsley piece. Woodhouse did however credit the two Yale professors for their groundbreaking work.

Predictably, the two reports differed. Where Silliman found some parts of the stone attracted by a magnet, Woodhouse did not. Silliman recorded more silex, more iron, more magnesium, less sulphur, and slightly more nickel than Woodhouse.[29] Addressing the elements contained in a mete-orite was the first step in convincing the public that they were delivered from outer space. Silliman's analysis established that the rocks contained iron, silex, magnesium, nickel, and sulphur. He also discovered that even the most intense heat couldn't evaporate these components.

It was Nathaniel Bowditch, a mathematician and astronomer, who authored one of the most controversial papers—controversial because it was President Thomas Jefferson who requested Bowditch undertake the independent investigation. Bowditch estimated that the Weston meteorite measured 491 feet across and weighed at least six million tons.[30]

"The extraordinary meteor which appeared at Weston in Connecticut on the Fourteenth of December 1807, and exploded with several dis-charges of stones, having excited great attention throughout the United States, and being one of those phenomena of which few exact observa-tions are to be found in the history of physical science, I have thought that a collection of the best observations of its appearance at different places, with the necessary deductions for determining as accurately as possible, the height, direction, velocity, and magnitude of the body, would not be unacceptable to the Academy," Bowditch wrote in his paper. "Since facts of this kind, besides being objects of great curiosity, may be useful in the investigation of the origin and nature of these meteors; and as the methods of making these calculations are not fully explained."[31]

Like Silliman, Bowditch concluded the sheer size and weight of the Weston meteorite indicated it came from outer space.

In 1810, Jeremiah Day pondered the question of the meteorite's origin a second time. "In what way then can they be carried up, fifty or a hun-dred miles, from the surface of Earth?" Day asked. "What is there in the atmosphere, which could give them their rapid horizontal velocity?"[32]

The answer: absolutely nothing. Nothing in Earth's atmosphere could produce such a rock. It was impossible for clouds to produce, let alone heat, meteoritic stones. "A solid substance elevated to a great height, and left to itself, would descend very rapidly. . . . Its velocity is such as could not be produced by the atmosphere," Day wrote as he further explored Silliman's thesis. "It must require a strong faith to believe that the atmosphere, even if furnished with materials, could produce such a body, and then give it a velocity sufficient to carry it beyond the circumference of Earth."[33]

These scientists built on Benjamin Silliman's work, as Silliman had built on the work of previous scientists. Nevertheless, it was Silliman who unequivocally succeeded in proving lunar volcanoes didn't jettison meteorites. He proved that the moon couldn't heave meteorites; doing so would in fact violate a principle of nature. The moon orbited Earth; therefore, anything a hypothetical lunar volcano exhaled would have to orbit Earth.

Benjamin Silliman and James Kingsley's report also laid the foundation for the future study of meteoritics. Everything scientists know today about meteorites—the different types, the various minerals, trace elements, and isotopes contained within—began in Silliman's laboratory. In the past two hundred years meteoritics has progressed significantly, despite numerous hiccups. Even as late as the 1940s scientists debated whether a volcano or an extraterrestrial impact created certain geological formations on Earth, like the meteor crater in Arizona.

Professor Benjamin Silliman made certain that his work spoke to those with religious concerns. He considered himself first and foremost a man of science; yet, he was also a devout Christian. Silliman championed the idea that God revealed himself through nature. Silliman balanced science with faith. His work, and the work of other early nineteenth-century natural philosophers, opened the way for American Transcendentalists such as Ralph Waldo Emerson and Henry David Thoreau, who believed that there could be harmony between the spiritual and the physical.

While Silliman labored on his thesis, religious groups attacked many geologists for teaching theories that contradicted the biblical account of creation. Silliman discovered that sparking a public interest in science was easier than avoiding the wrath of religious fundamentalists. At the same time, he recognized those who saw the omnipresence of God. "God [is] to be honored as the great author. . . . The nature of his works is to be learned by study and investigation," he wrote. "This Earth is the only heavenly body which we can examine. Meteorite stones contain the only notices of foreign bodies that have come to us."[34]

Silliman's intrinsic ability to reconcile the creation story in Genesis with the latest geological discoveries contributed to his success. Silliman seldom directly engaged in the debate over geology versus Genesis. He actively promoted the marriage of faith and science in his lectures. He often said that he believed Earth was created in an instant, but that it was subsequently subjected to a long course of formation. This gift won over many of his more religious contemporaries who might otherwise have rejected his findings. Even so, Silliman seldom directly engaged in the geology versus Genesis, or minerals versus Moses, debate. He had no intention of undermining the Bible. In 1833, he penned a supplement to the second American edition of *Consistency of the Discoveries of Modern Geology with the Sacred History of the Creation and the Deluge.* Here he sought harmony between divergent ideas.

> In this country, the cultivation of scientific geology is of so recent a date that many of our most intelligent and well educated people are strangers even to its elements, are unacquainted with its amazing store of facts, and are startled when any other geological epochs are spoken of than the creation and the deluge, recorded in the Pentateuch. . . . But, there is no reason to believe that any part of the crust of the earth, reaching even to a fathomless deep, is now in the condition in which it was originally made; every portion has been worked over and brought into new forms, and these changes have arisen from the action of those physical laws which the Creator established.[35]

Silliman understood that a portion of the populace would object to the fundamentals of science, no matter what. "Such a man can believe anything, with or without evidence. If there are nay such persons we leave them to their own reflections, since they cannot be influenced by reason and sound argument; with them we can sustain no discussion, for there is no common ground upon which we can meet."[36]

Until what came to be known as the Weston Fall crashed into the rocky soil of a small Connecticut town, most people scoffed at the notion that meteorites came from outer space. Silliman and the meteor became inextricably linked as word of the marvel spread throughout the nation.

The young professor had observed, tested, and recorded results that stood the test of time. Silliman himself said the Weston meteorite "was admitted to be one of the most extensive and best attested occurrences of the kind that has happened, and of which a record has been preserved." Hence he showed that the rough, black stones could have come only from outer space.

Chapter Seven

THUNDERSTONES

During the course of his lifetime, Benjamin Silliman filled many pages in his leather-bound journal with accounts of history-making objects falling from above. Many stories from antiquity intrigued him. Ultimately, these stories illuminated his muse when he tackled his own report on the December 14 fall.

One story that enchanted the Yale professor described the aftermath of a meteor fall on a Chinese village whose name is no longer known. The peasants whispered that the fall foretold of their emperor's death and the division of his kingdom. According to legend, fear so consumed the leader that he smashed the stone to smithereens. Still terrified, he massacred the people to quiet their murmurings. His actions failed. The emperor died the following year, and three years later his enemies divided his empire and extinguished his dynasty.[1]

Through his readings Silliman pondered how ancient cultures deeply believed that these otherworldly objects landed bearing messages from the gods. Some groups saw meteorites as harbingers of doom; others believed they foretold good tidings. Either way, most believed the stones possessed magical powers or at least contained quantities of valuable gold or silver.

Another story that intrigued Silliman occurred nearly four hundred years before the meteorite fell on Weston. On November 7, 1492, a solitary German boy stood in a wheat field. Suddenly, the midday sky above Ensisheim (now Alsace, France) turned eerily bright. Then it appeared to explode. The boy, whose name remains lost to the ages, ran straight to the village center. As the sole eyewitness he led the townspeople back to the field. They discovered a large stone buried in a three-foot-deep hole. The chief magistrate ordered the locals to haul the stone inside the church. In accordance with medieval custom, the people chained the rock to the wall with iron crampons, or grappling hooks, lest it fly away. Ignoring the sanctity of the place, villagers amassed inside bearing chisels and tools. They chipped pieces off the extraordinary mass and fashioned them into good-luck charms.

Some weeks later, Emperor Maximilian occasioned to travel through Ensisheim. He fervently believed the stone, weighing an estimated 260 pounds, was a portent of divine protection. He decreed that it symbolized God's intervention on behalf of Germany during its various conflicts with France. Maximilian ordered the meteorite to be forever preserved inside the church.[2] "Forever" ended during the French Revolution, when revolutionaries absconded with the stone. They dragged it to Colmar, France, and displayed it inside their own church. Today, a remaining fragment weighing 122 pounds resides in Ensisheim's town hall.[3]

Scientists consider Ensisheim the first eyewitnessed meteor fall in the Western Hemisphere for which preserved specimens still exist. However, most experts agree the oldest-surviving meteorite fell at Nogata, Japan, on May 19, 861 CE. Once resting inside a wooden box for a millennium, this fist-sized stone is now on display at the Suga Jinga Shinto shrine in Nogata, Japan.[4]

The oldest-known record of a meteorite fall of any sort dates back about four thousand years ago to Phrygia, now a part of west central Turkey. After the fall, a royal procession purportedly escorted the stone to Rome, where residents worshipped it for the next five centuries.[5] In 1807 came the Weston Fall—the first-recorded meteorite in the New World.

The legends about shooting stars have lived on for centuries. For instance, from Europe to Eurasia to North America, people believe wishing on a falling star guarantees that the wish will come true. People thought gods occasionally opened heaven's dome to spy on the small people of Earth as they tended to their daily affairs. Each time the heavens parted, the gods released a star. If one were lucky enough to wish upon a shooting star while its light radiated but before the star faded from view, the gods would hear and grant the wish.

While some saw falling stars as auspicious, others viewed them more ominously and assigned death and misfortune to the dazzling light. In William Shakespeare's *Julius Caesar*, Calpurnia, Julius Caesar's wife, said: "Fierce fiery warriors fight upon the clouds. The heavens themselves blaze forth the death of princes."[6] The tenth chapter of the Old Testament book of Joshua described how after the battle of Gibeon great stones were cast down from heaven during the flight of the Canaanites. The stones killed more men than did the swords of battle.[7] In 1513, Diebold Schilling painted a canvas depicting Halley's comet swirling amid a storm of flesh and blood. In 1664, John Danforth of Massachusetts preached that comets precede and "portend great calamities. . . . [R]epent for your sins and pay heed to this sign from an angry God." In 1680, the Puritan clergyman Cotton Mather echoed his words.[8]

Other groups interpreted meteorites with greater fatalism. For instance, in Lithuania people believed Fate started spinning the thread of each person's life at birth, sewing it to a star. At death the thread broke and the star fell. The Roman Catholic tradition linked shooting stars to human souls. Some believed shooting stars were actually wandering souls; others believed that they were souls ascending from purgatory to heaven. Some thought shooting stars represented suffering souls seeking prayers. If one uttered "rest in peace" three times before the meteor vanished, the soul would supposedly travel safely to heaven. As late as the 1800s, the notion that shooting stars ushered in evil lingered in isolated populations throughout Europe, Africa, and North America. In some parts of America, misfortune fell only if one pointed to the meteor.[9]

Countless examples of prophecies, speculations, and myths linked to meteorites fill Silliman's journal entries. He had learned of them during his study abroad. People in Mesopotamia believed meteorites were demons or devils. Some thought the glowing orbs were fiery dragons racing across the black sky. Even the myth of Kronos may actually be a legend about a meteorite. The Titan ruler ate his newborn children to prevent them from ever usurping his throne. According to legend, after Rhea gave birth to Zeus, she tricked her husband Kronos into swallowing a stone. Later, after Zeus conquered the Titans, Kronos regurgitated the stone, thought to be a meteorite, driving it straight to the center of the world.

Interpreting cosmic events continues to this day. In 1997, the Heaven's Gate cult committed mass suicide while the comet Hale-Bopp crossed Earth's orbit.

Today, a rock falling from the sky is weird and wonderful, whether the event is experienced by a scientific observer or an amateur eyewitness. But as recent as two centuries ago, mystery continued to shroud such events. Early civilizations understood meteorites came from space, but they explained them with myth and legend.

Many ancient philosophers supposed meteorites were other-worldly bodies that somehow loosened their planetary shackles before plunging to Earth. Some collected the debris and then incorporated findings into their history. Early writings indicate that Egyptians, Greeks, Romans, and the Chinese understood that meteorites fell from the sky. In 350 BCE, during the age of Aristotle, people recognized meteors as atmospheric phenomena. Indeed, the word shares the same root as the Greek word *meteorios*, or *meteora*, which means "things that are lifted high in the air."[10] Aristotle surmised that stones couldn't come from planets since that violated known and accepted laws of heavenly perfection. Ancient Greeks believed that pieces of heaven would never land on such an imperfect place as Earth. However, Aristotle also speculated that meteorites could not have been formed in the atmosphere. No, the philosopher claimed, Earth formed the stones deep within and then great winds heaved them into the air before hurling them back to the surface.[11]

Yet, this early awareness that the rocky masses were not of Earth vanished during the deep intellectual slumber of the Dark Ages. Thus these fascinating stories and folklore held Benjamin Silliman spellbound as he embarked on his career in chemistry and geology. They fueled his imagination and increased his hunger for more knowledge.

While researching meteorites and earthly rocks, Silliman learned people had long kept track of these extraterrestrial visitors. Recent scholarship of ancient texts reveals that the Hittites, an ancient people who dwelled in Syria about 3,200 years ago, understood that meteors came from space. The Hittites realized the stones yielded iron, naming iron *ku-an*. Some scientists consider this the earliest-known name for meteoritic iron.[12] Sumerian texts dating from 1900 BCE mention a precious sample of metallic iron that could have come only from a meteorite.

In the past several decades, archaeologists have uncovered many artifacts crafted from meteoritic iron such as knives, swords, and scimitars. The Greeks mention iron in their story of Achilles, and various Greco-Roman coins picture sacred stones on their stamped surfaces. It was said that at the funeral games of Patroclus, Achilles offered a mass of iron, most likely a meteorite, as a prize for his fallen friend.[13]

In 1853, the East African Wanika tribe dressed a fallen meteorite in ceremonial regalia before anointing its stony surface with sacred oil. Unfortunately, the ritual didn't halt their enemies from torching their village. The Wanikas decided that the newly arrived deity was a poor protector and promptly sold it to nearby missionaries. North American Indians also revered meteorites, often burying them in crypts. In Casas Grande, Mexico, archaeologists discovered one meteorite swaddled in mummy wrappings and carefully placed in the burial ground of Montezuma. Comanches living in Texas once laid a meteorite at a trail intersection so passersby could place beads and tobacco at its feet.[14]

While the Plymouth colonists were settling into their new home, a burning mass fell on the other side of the world in a village just southeast of Lahore, which was then part of India. The explosions were deafening and the grassy lands were singed, according to legend. The hot ground surprised the men who were digging up the stones. Finally, the village leader called upon a skillful artist to craft a saber, a knife, and a dagger from the stone's body. Legend insists the stones were so hot and brittle, they flew to pieces as the first hammer blow struck.[15] Over the centuries the idea that Earth spewed forth these stones faded, only to be replaced by the belief that clouds in Earth's atmosphere gave birth to them instead: "It is but a trite observation to say, that the clouds make frequent visits to the waters of Earth, from which they usually carry away large quantities of that element, and with it, no doubt, the substances (even some of the fish) which form the beds [are blended, and] it is that these substances may be concreted; and, by some extraordinary concussion in the atmosphere, return to Earth."[16]

With the arrival of the Renaissance in the fourteenth century, beliefs slightly changed. People started considering the world as it truly was, and not as they imagined or wished it to be. Europe's scientific community began noting how nature worked independent of the influence of personal beliefs.[17] For example, Galileo confirmed Copernicus's heliocentric world model while also discovering that planets were more than heavenly lights spinning along with Earth around the sun. His telescope revealed that not only were these orbs other worlds, but that some also had moons.

In 1758, nearly five decades before the Weston Fall, Halley's comet passed over North America. Thomas Clap, Yale's president at the time, saw it streak across the sky, as did John Winthrop of Harvard, and future Yale president Ezra Stiles of Newport, Rhode Island. Advocating a popular theory, Clap said meteors were large objects circling the planet in elliptical orbits, occasionally passing close enough to Earth to be visible. Nevertheless, Clap didn't understand that meteorites subsequently fell from the sky to the ground.[18]

Many people believed thunderstorms formed meteorites. They thought lightning fused together dust particles inside clouds. Growing too heavy, the clouds fell from the sky, hence the name *thunderstones*. At the dawn of the eighteenth century, the scientific community still hadn't witnessed a thunderstone form in the atmosphere. Moreover, most scientists persisted in arguing that stones couldn't fall from space. But then, just a few years later, on July 24, 1762, a streak of light raced across the sky above the village of Agen, which is situated between Bordeaux and Toulouse, France. Men, women, and children heard thunderous explosions. Then, after a pause, a shower of stones pummeled the countryside.

Still, the puzzle pieces didn't fit, though not for lack of trying. Ernst Chladni (1756–1827), a German physicist and lawyer, became one of the first people to raise the idea of extraterrestrial origin of meteors. He did this without ever laying eyes on an actual meteorite. Chladni, whom the church considered a heretic, believed these rocks had journeyed through space until Earth's gravity drew them into its atmosphere. Toward the end of the eighteenth century, when Enlightenment ideas dominated the European mind-set, Chladni revived the ancient idea that objects fell from the sky. He dared to gainsay the accepted wisdom, handed down from Aristotle and supported by Isaac Newton, "that no small bodies exist in space beyond the moon."[19]

Chladni couldn't understand the controversy. After all, ancient art depicts all sorts of objects raining down from space. He also argued against the claim that outer space was barren. He opposed commonly held ideas that clouds, lightning, or the aurora lit fireballs.

> The falling of stones from the clouds is an event which has frequently happened in Europe, in Asia, and in South America. The accounts of such phenomena were, for a long time, rejected by Philosophers as the offspring of ignorance and superstition. . . . It is now admitted not only that such phenomena have existed in modern times, but that the accounts of similar events in former ages are in a high degree probable.[20]

Nevertheless, for every scientist and citizen who started believing in more reasonable explanations, there were people, such as Pierre Berthollet, who clung to their views. "How sad it is that the entire municipality enters folktales upon an official record, presenting them as something actually seen, while they cannot be explained by physics nor anything reasonable," wrote Berthollet in 1791. Berthollet, who did not see any fall and didn't give weight to corroborated eyewitness accounts, wrote this even after the town of Barbotan, France, provided notarized testimony about a meteor fall.

In 1795, the peasants of Wold Cottage, England, witnessed an incredible sight. On December 13, the first-recorded meteorite to fall in Britain landed a mere nine yards from where John Shipley, a seventeen-year-old plowman, stood. Had it not been for Edward Topham, the event might have been relegated to the ranting of uneducated people. But Topham was a landowner, in addition to a soldier, a playwright, and a newspaper owner. His background gave him credibility. Like Silliman, Topham studied science at the University of Edinburgh. He recorded the testimony of those living on his land. In time, people flocked to see Topham's stone. He later donated the sizable stone to a museum.[21]

While Wold Cottage was the first of three successive falls that helped convince the world that meteorites came from outer space, people and scientists remained skeptical especially because another event convinced some that meteorites came from bodies orbiting Earth. On January 1, 1801, Giuseppe Piazzi (1746–1826) officially labeled Ceres, the largest, and first, dwarf planet in the asteroid belt, which had first been observed in the 1760s. It joined the moon as a suspected source of meteorites. And though some held that meteorites might come from the asteroid belt, others continued to theorize that meteorites were rocks thrown from lunar volcanoes. Others cleaved to the idea that terrestrial volcanoes jutting forth from icy lands near the North Pole jettisoned the rocks.[22]

Later Silliman would refer to these historic accounts when lecturing and submitting papers before scientific societies. In 1803, he had written in his journal about another remarkable case. A cloud near the horizon suddenly interrupted the clear and calm air of L'Aigle, France. A series of explosions peppered the air before a blizzard of stones rained down. Close to three thousand stones reportedly fell on this small French town; the largest stone weighed about 17.5 pounds. The stones rolled from roofs and cracked tree branches. They ricocheted off the hard ground and made the soil smoke. The "stones were so hot that one burned the hat of a peasant and those that took them up were obliged to throw them down. This meteor was very splendid even by full day light," wrote Silliman.[23] Terror struck the people; a laborer reportedly fell prostrate, pleading for forgiveness of all his sins.

L'Aigle merits attention as well because of the sheer number of witnesses who reported similar experiences. People from "all professions, manners and opinions, ecclesiastics, soldiers and laborers, men, women and children, agree in referring the event to the same day, the same hour, and the same minute."[24]

Indeed, the scientific revolution, in respect to meteorites, occurred in fits and starts. There was no sudden irreversible reordering of the way people viewed shooting stars. It would be some time before people knew that Earth plows through hundreds of tons of meteors every day. Orbiting in a belt between Mars and Jupiter, these meteors make shooting stars. In 1901 the *New York Times* reported: "An interesting result of the century has been the establishment of a general similarity between shooting stars and meteorites."[25]

In his day, Benjamin Silliman's views about meteorites ran counter not only to the astronomical attitudes of ancient civilizations but also to the more muddied ideas of his contemporaries. Silliman's theories contradicted set views.

Meteorites are the most ancient, primordial pieces of planetary matter people can touch and examine. Predating the most primitive rocks here on Earth, they were formed well before our planet was a smoldering satellite of the sun.[26] They are artifacts of a time when interstellar dust and gas swirled about the universe. They predate the birth of the sun and its clan of planets and satellites, asteroids, and comets.

Humans exist on the thin skin of a large body that travels through space. Studying meteorites offers crucial information about Earth's origins and its cosmic surroundings. Absent astronauts' efforts to hunt and gather rocks from space, meteorites remain the only physical samples of other bodies that also spiral through the galaxies. They are materials from beyond Earth that one can hold and study.[27] Scientists have unearthed fossilized meteorites as old as 475 million years in the Ordovician limestone in Sweden. Yet when most direct evidence of Earth's birth has been erased, meteorites are the volumes that record the solar system's first steps.[28]

Meteorites, including interplanetary dust particles, or IDPs, contain primary historic evidence of an early solar system history, from its formation to the composition of asteroids and comets. These stellar objects hold some of the best clues as to the nature of events that occurred some 4.5 billion years ago. Life likely began on Earth when giant meteorites from outer space crashed into what was then a frozen planet. They melted and heated vast seas just enough for the first life-forms to emerge.[29]

According to scientists, these meteorites unleashed energy ten million times greater than the nuclear bomb dropped on Hiroshima. The first meteorites increased the diversity of life since they killed off old ecosystems, creating space for new species to evolve. Some of the early craters on Earth's crust may even have served as natural beakers that brewed and stewed the new chemicals that contained essential ingredients in the recipe for life, becoming what Charles Darwin would refer to as the "warm little pond."[30]

Meteorites link the past and the future, outer space and Earth. During its formation, Earth likely experienced a heavy bombardment phase similar to that experienced by both the moon and Mercury. However, while the massive impact craters remain visible on those planets, layers of constant change hide those craters on our big blue planet. Evidence might lie under the detritus, but for now scientists must rely on meteorites, virtually unchanged since their formation, that survive their fiery descent. Meteorites salted Earth with organic matter—with life itself.[31]

In the last two hundred years more than 140 meteorites have fallen in the United States. These meteorites, which come from elliptical orbits in the asteroid belt of our solar system, are fragments of worlds that exploded long ago.

As meteors flash across the sky, a blue-white or reddish-yellow light frequently radiates from them. But the real drama arrives when the night is rendered as brilliant as a day. This happens when, unannounced, a large fireball, following a billowing, luminous trail, flits across the sky, all to the aural accompaniment of whizzing and the clatter of thunder. Lasting only a few spellbinding seconds, the display concludes with an explosive detonation. In contrast, a small meteorite may fall with little fanfare. It might simply appear as a fiery mass in the sky, traveling swiftly in an arc, before leaving a short-lived luminous trail behind.

Fireballs on the order of the Weston Fall populate a class by themselves because they are massive enough to survive the friction of our atmosphere as they descend. Disintegrating into several smaller pieces before reaching Earth's surface, these fireballs may be silent or they may produce violent detonations, as occurred in Weston.

Since most meteorites are between 17 and 20 percent iron and oxide, this gives these rocks a rusty color. They are heavier than they appear.

Iron meteorites may be the best-known meteorites, but they actually account for only 10 percent of all meteorites that reach Earth. Stone and stony-iron comprise the rest. While many scientists think that most iron meteorites originate from the metallic cores of planets or planetoids, some speculate exploding supernova stars can create them. Meteorite metal has a unique structure resulting from being melted in deep space and then crystallizing in the near absolute-zero high-vacuum void of outer space.

The most precious and valuable meteorites are of the stone variety because they are hard to distinguish from Earth rocks. These meteorites come in two classes: chondrites and achondrites. Weston's was a chondrite. Achondrite meteorites may be fragments of other mature planets, moons, or asteroids that traveled through the solar system for eons before landing on Earth.

Meteorites get their spherical or elliptical profile as they tumble through the atmosphere during the ablation process, that is, when the outer layer of the meteorite becomes vaporized. Their surface might be smooth or pitted. When meteorites pierce Earth's upper atmosphere, the friction simultaneously melts and removes their outer surface. This long plunge through the cold of our lower atmosphere cools the meteorite considerably and transforms the stony surface material into a glasslike substance.

Contrary to reports from L'Aigle, meteorites don't ignite grass or fall in flames. In fact, relatively little heat actually penetrates the stone's core. People who happen upon freshly fallen meteoroids find they aren't usually very warm to the touch. This poor heat conduction allows meteorites to survive in generally pristine condition. The folklore and legends were flawed when it comes to describing the stones as too hot to handle.

The glowing trail most witnesses reported during the Weston Fall resulted from deionization, the removal of atoms and ions. Most meteors are actually large dust particles (dust motes) or small fragments, including most of those that are seen in meteor showers. A meteorite's mass and velocity at the moment it enters Earth's upper atmosphere decide when people will first notice the burning orb. Usually, they appear about sixty miles above Earth. Here atmospheric density is strong enough to heat the meteoroid to incandescence. The atmospheric gases surrounding the heated mass make the fireball seem larger and closer than it really is. The light generated from the flight is usually visible for hundreds of miles from the spot where the meteorite comes to rest.

Witnesses also report echoing rumbles. Scientists now understand that the noise resembles a jet, which, when traveling at or above the speed of sound generates a pressure wave, or shock wave, in front of the plane. On the ground, the wave sounds like a sonic boom, which can be heard up to thirty miles away.[32]

Babylonian mythology explained stars as points of light suspended on strings that are pulled up in the daytime and let down at night. But in 1807 most Americans still believed rocks were the gifts of Earth. They existed to build stone walls, fire pits, or to be dug out so wagon wheels wouldn't trip over them. The last thing rocks were supposed to do was fly through the air. Should that happen, it would be a terrorizing spectacle, enough to make the stoutest heart quake. Yet these stories didn't scare Benjamin Silliman. They intrigued him. When he arrived in Weston to interview witnesses, he arrived ready to set the record straight.

Chapter 8

THE MISQUOTE HEARD ROUND THE WORLD

It is said that when President Thomas Jefferson first heard of the Weston meteorite, he proclaimed, "I would more easily believe that two Yankee professors would lie than that stones would fall from heaven."[1] In another version of the story, Jefferson and Samuel Latham Mitchell, a naturalist and politician, were dining together one evening. Mitchell later repeated their conversation from that night:

> My correspondents, Holley and Brunson, who went early on a tour of exploration, wrote me an account of their adventure, and sent me by the mail a specimen of the aerolithe. . . . The news excited a great sensation, particularly as the whim was then prevalent that the productions were ejected from the moon by volcanic fire. . . . [T]he curiosity of a senator who lodged at the same house with myself was worked to a high pitch. He had accepted an invitation to dine with the President that day, [and] he returned from the party indignant at the reception of his story. He said Jefferson produced the most perfect sang froid, or provoked a sort of scornful indifference.[2]

There is a third version of the president's reaction to the professor's work; in this account Thomas Jefferson conveyed his thoughts on the subject most succinctly: "It is all a lie."[3]

None of these statements can be proven true. However, more than two hundred years later the legend persists. Its lasting power is a testament to the era's turbulent economic, political, and social climate.

It was mere happenstance that the Weston meteor kited across the sky on December 14, 1807, the same day President Jefferson's Non-Importation Act, which restricted trade with Great Britain and France during the Napoleonic Wars, went into effect. Eight days later, on December 22, Congress passed the first Embargo Act, which closed all American ports to foreign trade. Although Benjamin Silliman attached no spiritual significance to the meteor fall, the timing of the two events was remarkable. For as a student of history, Silliman knew the ancients believed meteorites foreshadowed calamity; both pieces of legislation plunged New Haven, and much of New England, into catastrophic economic decline. Ports closed and vital maritime commerce diminished, both of which had long sustained the region's prosperity. The verbal attacks between the Federalists and the Democrat-Republicans intensified, and the political climate worsened. Passage of these two acts sank any chance for reconciliation between New England Federalists and Jefferson supporters. Although Jefferson was midway through his second term, reconciliation was important for the success of his political party.

Silliman and Jefferson were not cut from the same cloth. Silliman was a staunch Congregationalist who never supported the Virginian's presidency. The young professor associated with men who belonged to the Connecticut Wits, a loose-knit group of Yale students and rectors formed in the late eighteenth century. The group, whose members included Timothy Dwight and Silliman's future father-in-law, John Trumbull, devoted itself to the promotion of American literature as well as to the Federalist philosophy of a strong central government. The group also encouraged a heightened religious conscience, which pitted it firmly against Jeffersonian Deistic ideals.

For his part President Jefferson had little use for New England, Yale, or most anything Northern. Aside from the religious fundamentalism and its Federalist views, New England also irked Jefferson because of its zeal for commerce, its active expansion of the federal judiciary, and its alignment with bitter foes like Aaron Burr. This acrimonious relationship provided fertile ground for the seeds of a rumor to sprout. One of the earliest mentions of the quote is in "American Contributions to Chemistry: An Address Delivered on the Occasion of the Celebration of the Centennial of Chemistry," by Benjamin Silliman Jr., the professor's son.

The younger Silliman was speaking before an audience in Pennsylvania on August 1, 1874. He had come to speak on the centennial anniversary of the discovery of oxygen. He used the occasion to celebrate his father, Benjamin Silliman, who had died in 1864.

Born in 1816, Benjamin Jr. was one of four children. The marriage of his parents, Benjamin Silliman to Harriett Trumbull, the daughter of Connecticut's governor John Trumbull, ensured that politics received its fair share of attention during family discussions. Over the years the family exchanged views on everything from Jefferson's attitude regarding the meteorite to their strong abolitionist stance. As he matured into a scientist in his own right, Benjamin Silliman Jr. harbored resentment against the president who, in his eyes, snubbed his father's work.

Thus, Jefferson's decision to ask Nathaniel Bowditch to investigate the Weston meteorite stung Professor Benjamin Silliman, and the sting remained throughout his life. Although Jefferson didn't ridicule Silliman's work, it was interpreted as the president's lack of faith in Silliman's report. At the time, Bowditch's popularity far exceeded the growing popularity of Silliman. It riled his son, Benjamin Jr., because he thought Jefferson not only dismantled New England's commerce but also tried to usurp New England's discoveries by having Bowditch write a report that could have outshone his father's report. Aside from spending a lifetime devoted to chemistry, Benjamin Jr. spent quite a lot of time devoted to undermining Jefferson, for whom he harbored a deep political hatred.

Thomas Jefferson was well into his second term when the sky blazed over Weston, Connecticut. By then several crises had frequented the Oval Office. The Barbary Coast Pirate incident still dogged the president, and the recent Chesapeake Affair burdened him. Both incidents exposed America's vulnerability at sea and abroad, and so Jefferson wanted to end America's apparent inferiority. Both events affected New England, and both events led to the embargo.

After Great Britain and the United States signed the Treaty of Paris in 1783, America wasted no time in increasing its international trade. New

England ships bore much of that commerce, which in turn brought considerable profit to the merchants and the ship owners who hailed from that region. The wealth allowed the Northeast to prosper and invest in its thriving schools and towns.[4] However, Britain and France remained embroiled in the Napoleonic Wars. Officially neutral, the United States attempted to maintain its increasingly lucrative trade with the belligerents, but it proved nearly impossible. American shipping grew ever more vulnerable as both powers sought to crush the economic life out of one another. Their strategy included attacking American vessels. England and France frequently seized the vessels of neutral nations if they believed the ships carried cargo destined for their enemy's ports. However, Britain took matters one step further, frequently plucking seamen from American ships and conscripting them into the Royal Navy. Most of these sailors professed US citizenship. Constantly hunting would-be deserters, the Royal Navy usually disregarded their claims. This policy nearly led to war with America.

On June 21, 1807, six months before the meteorite hit the ground, the British warship HMS *Leopard* fired upon the American warship USS *Chesapeake*. The incident occurred off the coast of Norfolk, Virginia. Commodore James Barron commanded the *Chesapeake*, while Salisbury Pryce Humphreys commanded the *Leopard*. Humphreys requested permission to board the American warship, ostensibly in search of suspected deserters from the Royal Navy. As expected, Commodore Barron refused. In an angry retort, the *Leopard* fired broadsides, killing three Americans and wounding eighteen. One of the injured, Robert Macdonald, died after shipmates carried him ashore.

The British boarding party found four suspected deserters hidden among the *Chesapeake*'s crew. The Royal Navy seamen were identified as David Martin, John Strachan, and William Ware from the HMS *Melampus*, and Jenkin Ratford, who had served on the HMS *Halifax*. Ratford was the sole British-born subject among the group; all of the others were American born. After their arrest, Commander Humphreys transported the prisoners to Halifax, Nova Scotia, to await trial.

Fortunately for the Americans, the British judge commuted their sentence of five hundred lashes each. Such luck eluded Ratford, who was put to death. Officers promptly hanged Ratford on the yardarm of the *Halifax*. Eventually, the British government returned the American men to the United States and even paid reparations. Yet President Jefferson considered the gesture to be too little and too late. He thought both Great Britain and France violated US sovereignty in ways that suggested they still considered America to be a mere colony. Jefferson, who was already struggling with efforts to manage the vast landmass of the recently purchased Louisiana Territory, aimed to secure the new nation's rights on both sea and land. He ordered all British ships to vacate American sovereign waters immediately.

Jefferson convinced Congress to pass the Embargo Act, a piece of legislation intended to impress foreign powers with America's commercial might, but in fact it targeted American shippers and their vessels. Many New Englanders subsequently condemned the new law as antitrade because no American vessel could dock in foreign ports without the president's personal permission. Furthermore, ships had to post a bond of guarantee equal to the value of both the craft and its cargo. The new law allowed minor coastal trading only if the ship's captain posted a bond for twice the amount of the merchandise's value. The government repaid the bond when the ship unloaded at another American port.

In the eyes of many, Jefferson appeared to unilaterally enlist American commerce in his battle to win recognition for America's sovereignty on the high seas. He thought that banning all trade, rather than resorting to guns and cannons, could realize that aim. For Jefferson and his supporters, the act could prevent the seizure of the nation's maritime resources and make Britain and France realize the futility of seizing ships sailing under the Stars and Stripes. Between 1803 and 1807, Britain seized 528 ships, while France captured 206.[5] The Jefferson administration wagered that Britain would quickly acquiesce since it relied so heavily on American raw materials for its manufacturing.

The Embargo Act didn't yield the desired result on the domestic front largely because the notion of economic self-sacrifice completely baffled Americans. Most people actually favored war, even if that meant losing their lives rather than their livelihood. "A state of war will always be deprecated by all wise and good men; but when it becomes unavoidable, a patriotic mind will cheerfully submit to all its calamities, rather than see the rights of his country wantonly violated," read an editorial in the *Boston Independent Chronicle*.

Thomas Jefferson's decision split the nation along stark party lines similar to the election of 1800. Again, newspapers engaged in scurrilous attacks, and editorial excess run rampant was the order of the day. No attack on a public person's character was considered too vulgar. Editors made sport of nearly every government decision; the embargo was no exception. In this way the nation's newspapers contributed to the debate and helped nourish the storyline of the Yankees versus the Virginian.[6]

"Under such circumstances the best to be done is what has been done; a dignified retirement within ourselves; a watchful preservation of our resources; and a demonstration to the world that we possess a virtue and a patriotism which can take any shape that will best suit the occasion," ran an editorial in the *National Intelligencer* on December 23, 1807.[7]

Other papers, such as the *New York Evening Post*, took an entirely different stand: "It requires all the confidence—all the faith, of which a stupid party bigotry is capable to approve of this terrible desolation [the embargo]—the justification of this dreadful butchery of the political body, requires indeed the sacrifice of all the pride, all the liberty, and all the good sense of the nation."[8]

Before the embargo became law, Jefferson and his team generated advance publicity in the nation's newspapers. "A crisis has arrived that calls for some decided step," they claimed on Christmas Day 1807. "The national spirit is up. That spirit is invaluable . . . the crisis not requiring war, still hoping, if not expecting peace, an embargo is the next best measure for maintaining the national tone. It will arm the nation. It will do more. It will arm the Executive government."[9]

The next day, December 26, the *New York Evening Post* published an article that stated in part that at "the present moment of uncertainty, apprehension, dismay and distress, every one is running eagerly to his neighbor to inquire after information. The measure at last adopted by the Government, of an Embargo, brings with it immediate bankruptcy to the merchant, and of course less of employ and want of bread to some thousands of laboring poor of this city."[10]

Aside from the newspaper campaign, Jefferson did little to win New England's understanding, let alone its support. He carried a grudge against Connecticut for opposing his reelection bid in 1804. Yet, because he had been overwhelmingly reelected, he figured that he had political capital to burn. As it happens, he didn't. The animosity with his political opponents intensified.[11]

Professor Benjamin Silliman made virtually no references in his diaries or letters to the politics of this tumultuous time, save for occasional musings about the looming war with Great Britain. Chemistry and geology consumed his attention. Meanwhile, on the Yale College campus, Timothy Dwight railed against Jefferson. He argued that the economic downturn would threaten his institution. Financial ruin visited cities and towns all along the seaboard; depression, unemployment, and unrest grew. The Federalists viewed the embargo as a costly, futile, partisan romp, entered into at their expense.

As a self-proclaimed neutral nation, the United States suffered under the embargo far more than either England or France. As the owner of more than half of the country's ships, New England's citizens suffered immeasurably. By 1805, New England owned half of America's shipping fleet. After the 1785–1789 depression, the United States enjoyed a time of real prosperity. Until 1807 New Haven's shipping industry reveled in its commercial success.[12] When the embargo was imposed its impact became abundantly clear after just a few weeks. Ships laden with goods clogged New England harbors. The smell of rotting meat, decaying wood, and putrid produce filled the air. In 1808, just one year after the act became law, Massachusetts's exports had sunk to just $5.1 million from its peak of $201 million. Moreover, shipbuilding tonnage constructed in 1808 was about one-third of that in 1807.[13]

The town of Salem, Massachusetts, one of America's prime commercial centers, suffered enormously from the fallout. During the first half of 1807, before the embargo, 134 vessels sailed for overseas ports. In 1808, not one ship cleared Salem's harbor. The next year, the town's soup kitchen nourished 1,200 destitute people. So desperate were those who owned and worked on America's shipping fleet, they volunteered to leave on British ships so that they might find work in foreign ports.

Moreover, scores of merchants and sailors evaded the new law. Several local governments in New England actually encouraged smuggling. Some refused to help the federal government enforce the embargo. On the Canadian frontier, especially in upstate New York and Vermont, people flagrantly violated embargo laws with smuggling routes flourishing along the shores of Lake Champlain. Breaking the law became so commonplace that federal officials considered the part of Maine that bordered on British-held New Brunswick to be in open rebellion. By March 1808, Jefferson's tolerance had run out. Resolving to enforce the embargo to the letter, he dispatched Capt. Stephen Decatur, a hero of the Barbary War, to cruise the New England coast and enforce the embargo. This so incensed New Englanders that Federalist politicians in the Nutmeg State rallied; the Federalists enjoyed a brief return to office. Although Federalists had kept positions in state government, they had lost congressional seats during Jefferson's administration.

"Jefferson's Embargo is an excellent mother, for she brings forth federal children in abundance," wrote a Connecticut correspondent in the September 8, 1808, issue of the *New York Evening Post*.[14]

Across the region Federalist editors were unabashedly abusive: "I wish that Jefferson and Simon Snyder [governor of Pennsylvania and Jefferson ally] were both in hell and a clog of brimstone at each foot!"[15] Even a song popularized this sentiment:

> Our ships, all in motion,
> Once whitened the ocean,
>   They sailed and returned with a cargo;
> Now doom'd to decay,
>   They have fallen a prey
>   To Jefferson, worms, and Embargo.[16]

Children even got in on the act. William Cullen Bryant, who would later gain fame as a distinguished poet and editor, was a thirteen-year-old boy living in Cummington, Massachusetts, when he penned this insulting poem.

> Go, wretch, resign the presidential chair,
>   Disclose thy secret measures foul or fair,

Go, search, with curious eye, for horned frogs,
Mongst the wild wastes of Louisiana bogs;
Or where Ohio rolls his turbid stream,
Dig for huge bones, thy glory and thy theme;
Go scan, Philosophist, thy Sally's charms,
And sink supinely in her sable arms;
But quit to abler hands the helm of state,
Nor image ruin on thy country's fate.[17]

In 1809, Benjamin Silliman's father-in-law, Connecticut governor Jonathan Trumbull, refused to employ anyone who openly supported Jefferson or the embargo. On February 4, 1809, the governor published a broadside announcing his intention. In addition, the Connecticut General Assembly declared the embargo "arbitrary, oppressive, and unconstitutional."[18]

While Jefferson occupied the White House, some political leaders, especially John Adams, thought it best to befriend the English Crown. Americans still depended on Great Britain for imported manufactured goods. Although American exports had increased—the first jump came between 1787 and 1788 and then again in 1792—it would be years before the former colonies produced their own finished products.[19]

On the other hand, Jefferson had spent several years posted to the French court. He favored France and the ideals of the French Revolution. His affinity for France raised suspicions among Federalists; they believed Thomas Jefferson was Napoleon Bonaparte's marionette. So convinced was Timothy Pickering, a politician from Massachusetts, that he announced a conspiracy between Thomas Jefferson and Napoleon Bonaparte to maintain the Embargo of 1807.[20] No plot existed, but the rumors illustrate the depths of the distrust and ill will. The citizens of New England felt Jefferson had singled them out for ruin in order to spare the rest of the nation from the cost of war.

As Benjamin Silliman steadily worked on his report, concerning the scientific breakthrough that the Weston meteorite presented the nation and the world, the people of Weston and the surrounding region waited in vain for desperately needed government support. New England Federalists contemplated delivering Congress an ultimatum: rescind the act or the region would withdraw from the Union. Murmurs of a split brewed. Although the Hartford Convention wouldn't meet until 1815, the Federalist Party flirted with the secessionist schemes of Timothy Pickering and Fisher Ames. Jefferson toyed with sending troops north to squelch secessionist talk in Hartford. Events in the state precluded any need for presidential action. The Hartford Convention permanently tainted the Federalists, and by 1820 the party became virtually extinct.[21] They were seen as disconnected from the rest of the nation and too married to upper-class interests.

Most historians agree that Thomas Jefferson stood for science and greater learning. His *Notes on the State of Virginia* provides an example of political and scientific eloquence that illustrated his varying pursuits.[22] In 1809, as he readied to leave the White House after eight tumultuous years, Jefferson said: "Nature intended me for the tranquil pursuits of science by rendering them my supreme delight."[23] Further evidence of his ardor dates to 1797 when he traveled to Washington to take the vice presidential oath of office. A wagonload of bones trailed behind his carriage. He brought them to the American Philosophical Society in Philadelphia, where he presented a formal research paper titled "A Memoir of the Discovery of Certain Bones of an Unknown Quadruped of the Clawed Kind, in the Western Part of Virginia."[24]

His fascination with fossils annoyed his political rivals. They thought he too easily neglected his proper, presidential duties. In 1808, when debate about the embargo intensified, another wagon delivered numerous specimens to Pennsylvania Avenue. Jefferson laid the fossils out in the East Room. His adversaries soon nicknamed him "Mr. Mammoth." As president Jefferson didn't embrace every facet of science. On the contrary, he dismissed geology as "too idle to be worth an hour of any man's life."[25] Clearly Jefferson and Professor Benjamin Silliman viewed science and its application differently. Initially, Jefferson saw science as practical, whereas Silliman saw it as more theoretical. Jefferson definitely favored botany and chemistry. He believed these disciplines could be harnessed for practical use in ways that geology and astronomy could not. Nonetheless, the two men desired that the new nation profit, intellectually and ultimately economically, with the help of science.

It seems that Jefferson intensified his interest in astronomy years later, well after founding the University of Virginia in 1819. He wanted Nathaniel Bowditch to take the position of professor of mathematics and astronomy, but Bowditch was committed to staying in Salem, Massachusetts. Later, in 1820, Jefferson would also work on creating an observatory. He had figured that it was important to make astronomical observations to fix boundary points on the Louisiana Purchase.[26] Land surveys based on astronomy were still used. He once wrote to John Adams, "I am indebted to Mr. Bowditch's very learned mathematical papers, the calculations of which are not for every reader, although their results are readily enough understood . . . the eccentricities of the planets of our system could oscillate only within narrow limits, and therefore could authorize no inference that the system must, by its own laws, come one day to an end."[27] The letter highlights Jefferson's discomfort with astronomy.

All the same, the Weston Fall piqued the interest of the third president. Aside from politics, Jefferson kept abreast of the latest trends in architecture, farming, and science. In light of this, Jefferson certainly must have known about European reports of meteorites.

Two weeks after the meteorite slammed into Connecticut farmland, Thomas Jefferson expressed his thoughts on the event in a lengthy letter to Daniel Salmon of Connecticut. Some critics read doubt between the lines of this letter, inferring that Jefferson doubted whether the meteor struck. "A cautious mind will weigh the opposition of the phenomenon to everything hitherto observed, the strength of the testimony by which it is supported, and the error and misconceptions to which even our senses are liable," Jefferson wrote. "It may be difficult to explain how the stone you possess came into the position in which it was found. But is it easier to explain how it got into the clouds from whence it is supposed to have fallen? The actual fact however is the thing to be established."[28] This apparent doubt helps explain why the quote about two Yankee professors withstood the test of time.

Whether the quote was real, a myth, or simply a burdened president throwing a barb at two Yankee professors isn't all that important, but what is significant is that Jefferson's views reflect the mind-set of a nation struggling to reconcile objective observation with deeply entrenched beliefs. This is precisely what Professor Benjamin Silliman worked to overturn.[29]

In 1807 most rational minds still couldn't grasp the notion that objects from space fell to Earth. Even Thomas Jefferson, the man behind the Lewis and Clark expedition, did not fully realize the vast expanse of the sky above. The president wasn't sure whether meteorites hailed from space or if they were the rocky offspring of atmospheric cloud formations. After all, most scientists of Jefferson's era considered phenomena such as meteors and fireballs unrelated to more legitimate astronomical events such as asteroids and comets. It took more than two centuries of painstaking research before scientists concluded that these occurrences are related.

Nevertheless, Jefferson was a capable scientist. His fieldwork in paleontology and agronomy met the highest standards. He read and wrote in Latin, Greek, Spanish, and French. In architecture, botany, and mechanics his vision surpassed most.

It might be said that Thomas Jefferson's passion for plants and his revolutionary outlook outweighed any lack of interest in outer space. In his mind, he was a farmer first and foremost, often stating it as his true occupation. Jefferson ceaselessly spoke of retiring from politics; he relished a return to nature. He loved riding through his plantation and walking among its gardens.[30] Always interested in new crops or the latest inventions in machinery, Jefferson was truly progressive. He believed that educated yeoman farmers would maintain a vibrant and strong democracy.

Like Benjamin Silliman, Thomas Jefferson appreciated the connection between the power of science and its possible contributions to society's well-being. When not serving in government, Jefferson tackled various methods and inventions. He worked to improve farming techniques from promoting crop rotation to reinventing the plow according to the scientific principles of Sir Isaac Newton. He developed methods for excavating archaeological sites that are still followed today.

Nevertheless, as has been noted, he actually disdained geology, judging it to have only limited usefulness. He "could not see any practical importance in knowing whether Earth was 6,000 or 6,000,000 years old, and the different formations were of no consequence so long as they were composed of coal, iron, or other useful materials."[31] In other words, the Weston Fall was nothing more than an interesting diversion. His disinterest in geology and mineralogy conflicted sharply with Silliman's view that the community's base of knowledge would be enhanced through a scientific understanding of the Weston Fall.

In 1803, four years before the fall, questions about meteorite origins reached the shores of America. Andrew Ellicott, a surveyor, mathematician, and friend of Jefferson, penned a letter to Robert Livingston, the US minister to France. He noted his reservations regarding the fall of stones from the sky. Livingston responded that scientists were vigorously debating whether meteorites came from volcanic eruptions on the moon or whether they were formed in the atmosphere. "As much remains to be said on both sides; prudent men have not yet thought it proper to pronounce judgment," he wrote Ellicott.[32]

Then in October of 1805, while Silliman was enmeshed in his studies overseas, Thomas Jefferson exchanged this letter with Andrew Ellicott about science, particularly about chemistry and meteorites:

I do not know that this would be against the laws of nature and therefore I do not say it is impossible; but as it is so much unlike any operation of nature we have ever seen it requires testimony proportionately strong. The formation of hail in the atmosphere is entirely unaccountable, yet we have the evidence of our own senses to the fact and therefore we must believe it. A most respectable sensible and truth speaking friend of mine gave me a circumstantial account of a rain of fish to which he was an eyewitness. I knew him to be incapable of speaking an untruth. How he could be deceived in such a fact was as difficult for me to account for, as how the fact should happen. I therefore prevailed on my own mind to sojourn the decision of the question till new rains of fish should take place to confirm it.[33]

A few months after the fall, on February 15, 1808, a Connecticut citizen wrote President Jefferson and offered him a fragment of the stone. He wanted the president to announce an official congressional examination. Jefferson turned down the request. Instead the president requested that Nathaniel Bowditch of Salem investigate the matter on his behalf. In effect, Bowditch retraced Silliman's steps and collected accounts from the many witnesses. He traveled to Wenham, Massachusetts, and interviewed Mrs. Gardner, who watched the fireball from her bedroom window. His investigation determined height, direction, and velocity. In the end it confirmed Silliman's work.

"The greatness of the mass of the Weston meteor does not accord either with the supposition of its having been formed in our atmosphere, or projected from a volcano of Earth or moon, and the striking, uniformity of all the masses that have fallen at different places and times (which indicates a common origin) does not, if we reason from the analogy of the planetary system, altogether agree with the supposition that such bodies are satellites of Earth," wrote Bowditch.[34]

President Jefferson pondered the question of meteorites and their parent bodies. But his doubt, coupled with his disdain for Yale, stemming from Dwight's rampages against him to his view that it was a fortress of federalism, as well as his disdain for Connecticut, prevented him from wholly welcoming Silliman's report. Silliman, on the other hand, never tried to personally educate Jefferson. However, Jefferson's view that astronomy wasn't important symbolized some of the resistance Silliman met as he worked to educate the American population on the importance of meteors and astronomy. Thus endured the tension between the professor and the president.

Chapter Nine

ROMANCING THE STONES

In 1810, workers punched through a dormitory room on the second floor of Yale College's South Middle Hall. Benjamin Silliman finally had an airy space, forty by eighteen feet, in which to house a mineral gallery. It would be a splendid room, scintillating with rare finds. He savored the time it took to carefully unpack his specimens, including a sizable piece of the Weston meteorite. He laid each one gently on its assigned shelf.

The specimen cabinet epitomized Benjamin Silliman's continued foray into the realm of public educator and science promoter. Science needed public interest and support to flourish, and Silliman's impeccable oratory and writing skills ensured this would come to pass. He would tell his audience about the winking lights in the night sky, and how he regarded stars as the gatekeepers to planetary systems other than our own.

"Nothing can exceed the splendor and magnificence of the starry heavens in a cold and cloudless night," he wrote. "It is, without doubt, owing to this cause, that occasional fiery appearances in the heavens, or in the atmosphere produces a far stronger impression than all the regular glories of the azure canopy."[1] Much later in life he realized that he had succeeded in reaching the general public. "And while I enjoy the satisfactory assurance that I have popularized science, these efforts brought important assistance to my family at a period when my children were requiring aid in their settlement in life."[2]

Indeed, just as his father, Gold Selleck Silliman, served the nation in war, Silliman served the nation in academia. This second-floor gallery helped to fuel America's mounting appetite for science even as the distracting stirrings of another war with Britain grew louder.

In Weston, a farmer with the last name of Jennings discovered a sizable piece of the Weston meteorite on his property. The tenacious farmer tried to sell a fragment of the Weston meteorite in New York City for the considerable sum of five hundred dollars. To ensure his fragment sold somewhere, Jennings tried to tempt Connecticut officials to bid on the stone as well, saying he preferred to sell it to the state government for the desired amount. Yale was, of course, interested in the purchase, but Jennings thought he might get more money from the state.

Upon hearing of the pending sale, two of Silliman's acquaintances entered the negotiations: family friend David Judson, whom Silliman often visited when traveling through Weston, and Dr. Isaac Bronson, who was riding home inside his carriage the day the meteorite fell. The two hoped to acquire the 36.5-pound fragment and wrote to Silliman of their efforts to persuade Jennings.

"Since I had the pleasure to see you another stone has been found near Tashua Hill, the most of any we have yet seen weighing 36.6 pounds—It is now in . . . the possession of a Mr. Jennings, and it will require some [text unclear] to get it from him," wrote Isaac Bronson. "I have offered him five [hundred] dollars if he would present it in his own name to your institution. I have requested your friends here to lay all the motives before him, which they can think of to induce him to do it. Jennings seems to entertain extravagant ideas of the value of his prize, and I doubt if he will be induced at present or any occasion to terms to part with it; but if it can be had you may rest assured."[3]

David Judson also tried to entice Jennings with the merits of selling the stone to Yale College. He told Jennings that the college would engrave Jennings's name on a plate fixed above the rock. Judson figured such recognition might outweigh coins in the coffer. Then Isaac Bronson tried his hand again, imploring the farmer to postpone selling the extraterrestrial prize. Their entreaties failed. Jennings took his meteorite fragments to New York City, where people queued up, paying a small sum to get a glimpse of the famous rock. It seemed Jennings was going to "put it, or keep [it] locked up in his chest twelve or thirteen months until he hears every offer that may be made from Boston to City Washington," Bronson told Silliman.[4] Eventually, the rock went to Yale.

As Americans settled lands to the west, farmers and prospectors found and sold many iron meteorites. Most of these finds were sold abroad, since for the greater part of the 1800s, American museums rarely mounted public exhibits geared toward scientific study. Even the Smithsonian Institution in Washington, DC, displayed little interest in meteorites until the 1880s. Ordinary individual collectors were rare, too. Instead, wealthy individuals assembled most of the nation's mineral collections. Mineral dealers opened curio shops in European cities, and agents scurried around the world searching for rare mineral specimens. Explorers returned from deserts and plains with loads of samples. Private collectors often exchanged minerals and presented rare specimens as gifts to friends and associates. Eventually, these private collections triggered the idea of recovering meteorites for scientific study, not just for artful display.

Today, meteorites are a popular collectible; museum-quality fragments command thousands of dollars in the marketplace. Collectors often pay handsome sums, from one to two thousand dollars per gram (about 1/28th of an ounce).[5] Until recently, the aesthetic aspects of collection didn't interest private collectors. Weight alone determined value. Now collectors search for stones that resemble something sculpted in the atelier of Alberto Giacometti or for historically significant pieces, such as those harvested near Weston. The latter have increased value for several reasons: they are bits of the first-recorded meteorite to fall in the New World; their origin was speculated upon in the so-called Thomas Jefferson quote; and because of their role in establishing the science of meteoritics.[6] Aesthetically pleasing specimens, or those with a rare mineral composition, can fetch more than five hundred dollars per gram. Meteorites from Mars, of which only twelve are known to have been recovered, sell for more than a thousand dollars a gram.

Today meteorites garner high prices because they remind people of how vulnerable they are to celestial objects passing so close to Earth. However, attaching a price tag to a celestial body has unique challenges. Determining ownership of a stone can be tricky. Sometimes lawsuits are required to resolve the matter. If a meteorite lands on private land, the owner has clear claim to the stone, no matter who found it. However, if the stone falls on public land, the governing body having jurisdiction owns it.

Nearly two centuries after the Weston meteorite fell, during a period between the mid-to-late 1990s, meteorite collectors vigorously competed for the limited amounts of meteoritic material appearing on the commercial market. Scientists and collectors alike coveted the most rare meteorite types. Discovery of Martian meteorites plus media coverage of the Pathfinder probe landing on Mars in 1996 caused meteorite mania to sweep the nation. Although the enthusiasm sometimes priced scientists well out of the market, it also gave financial incentive to make public the existence of rocks that might otherwise have remained hidden.

After the Weston Fall, the notion that rare rocks were solely the purview of the privileged began to change. A newfound respect for science helped people to become more observant of their surroundings, but the toll Thomas Jefferson's embargo had on American commerce also contributed to shifting attitudes. The clampdown forced many to seek new sources of revenue. More and more entrepreneurs literally went underground for riches, inciting a "metallurgic mania" in New England. Yet, because countless businesspeople lacked adequate expert information about the quantity and nature of the mineral ores, they made terrible investments in their mining endeavors.[7] To help mitigate these metallurgical misadventures, investors and mine operators began to educate themselves about the importance of metallurgic classification. Prospectors needed to understand what they had found. Slowly, metallurgy became a means to enrichment. The notion of trading and bidding on selected gems, stones, and pieces of rock falling from space had become a reality.

In the weeks after Silliman and Kingsley left Weston during December of 1807, a quiet network of collectors and dealers sprang forth looking to sell specimens to scientists, curators, and museums. Some yearned to possess the unusual, others dreamed of hefty profits for their sale. Many of these people were simple farmers like Jennings. Others were trappers and hunters who stumbled on meteorites while plowing fields or laying traps. Men like Silliman, Bronson, and Judson were relieved when many of those who found fragments of meteorites agreed to turn them over to museums or universities in the interest of science.

During these early years, curators sometimes traded meteorite specimens with dealers, collectors, and other museums. On other occasions a consortium of institutions might purchase a stone and then have it chiseled into several pieces so each could have a piece to study and display. This helped develop the science of mineralogy, which accelerated during the nineteenth century. It also allowed many more people, both in the public and in academia, the opportunity to see such stones. Scientists and academics increasingly convinced wealthy people to purchase rocks and gems and then donate their purchases to museums out of a sense of civic duty and to further scientific research.

Benjamin Silliman had an eye for minerals long before the meteorite fell. In 1805 he learned of Col. George Gibbs of Newport, Rhode Island. Gibbs's father had amassed the family's wealth through his mercantile business. When the elder Gibbs died, he left a sizable inheritance for his son George, who was expected to help maintain the family fortune. But commerce bored George, so like many wealthy young men of the period, Gibbs embarked on a grand tour of Europe. He had a passion for science, which he indulged throughout his extensive travels. For a time, George Gibbs moved to Lausanne, Switzerland, a hilly town overlooking Lake Geneva. There he studied under the tutelage of Heinrich Struve of Vaud, a mineralogist and expert in chemistry and natural history. Gibbs fell in love with minerals and gems. Soon he began collecting pieces, and in time his purchases greatly benefited American science. Gibbs traveled to Paris, where he bought a mineral cabinet from the estate of Jean Baptiste Francois Gigot d'Orcy. As the Inspector of Mines and Receiver General of Finances during the reign of Louis XVI, d'Orcy was also a patron of natural history. He amassed more than four thousand specimens during his forty years of collecting. D'Orcy had been guillotined during the Reign of Terror and his collection put up for sale. The acquisition of this amazing collection earned Col. George Gibbs a place as one of America's leading collectors of minerals. Benjamin Silliman yearned to see and handle the assorted specimens.

"This gentleman, as I was informed, had in 1805, brought over from Europe a splendid collection of minerals, augmented from time to time by magnificent additions, only a part of which has as yet been opened," recalled Silliman in his journals.[8] In 1808 the journal *Medical Repository* reviewed the cabinet:

> This rich collection consists of the cabinets possessed by the late Mons. Gigot D'Orcy, of Paris, and the Count Gregoire de Razamowsky, a Russian nobleman, long resident in Switzerland. To which the present proprietor has added a number, either by himself on the spot, or purchased in different parts of Europe. . . . In giving this account of a collection, so much wished for in the United States, it may be justly acknowledged, that it is principally by the assistance of the scavenges of France, that it was rendered so complete.[9]

A great variety of marbles, chalcedonies, and agates as well as gold and copper ores, chromates of lead, and jaspers now filled Gibbs's collection. In addition, newly discovered minerals, including spinel and oriental rubies, lined the shelves. All told it stood out as one of the largest geological collections, consisting of nearly twenty thousand samples. Before he left Europe to return home to Newport, Gibbs generously gifted many books, specimens, models, and instruments to the French Museum's National d'Histoire Naturelle and Ecole des Mines.

Before Gibbs's 1807 homecoming, Benjamin Silliman met the colonel's sister Ruth Gibbs in her Rhode Island home. After receiving her brother's permission, she invited Silliman to examine some of the minerals stored in a warehouse next to Gibbs's Newport mansion. When Gibbs himself returned stateside after the Weston Fall, he had his mineral collection shipped from Europe.

Finally, the professor and the collector met and forged a close friendship. Together the pair explored the mineralogy and geology of Rhode

Island. Silliman also helped his new friend unpack the many specimen boxes as they arrived from overseas.

"My daily visits to the Gibbs house, which was always accessible to me, made me familiar with its contents, and placed me on terms of easy interest with its liberal-minded proprietor. . . . I had now acquired a scientific friend and a professional instructor and guide, much to my satisfaction, and he appeared equally pleased to find a companion in his scientific sympathies and pursuits especially in a young man full of zeal, and both willing and desirous to work," Silliman recalled.[10]

Silliman hoped Yale would eventually acquire Gibbs's vast mineral cabinet, but he waited to ask the school to make the purchase. After having asked the Corporation to stretch the budget so frequently in the past few years, he didn't think they would agree to another sizable expense. As for Gibbs, he wasn't sure where he wanted to house his vast collection. He wanted to ensure that it went "in connection with some public institution or at least in such a position in some city that it could be made available to the promotion of mineralogy and of the connected arts and sciences."[11]

Gibbs offered the federal government the chance to open the collection at the Military Academy at West Point or even in Washington, DC, but the government declined his offer. After the refusal, Gibbs decided that neither schools nor museums in New York or Philadelphia were suitable hosts for his collection. Thus the boxes remained at Newport. Finally, during the winter of 1810, Col. George Gibbs resolved the matter of their disposition.

"Colonel Gibbs, on a journey, called on me in the evening, and, as usual when we met, the conversation turned on the cabinet, and I inquired: 'Have you yet determined where you will open your collection?' To my great surprise he immediately replied; 'I will open it here in Yale College, if you will fit up rooms for its reception.' . . . I lost no time . . . in laying the subject before President Dwight."[12]

Silliman arranged for Gibbs to loan his collection to Yale. The professor also persuaded the trustees to provide spacious accommodations for the pieces. By 1812, "The fame of the cabinet was now blazoned through the land, and attracted increasing numbers of visitors. I had become a zealous student of mineralogy and geology, and now felt that the time had come to present them with more strength and fullness than in former years," Silliman wrote.[13]

More than a decade later, Gibbs finally sold his entire collection to Yale for twenty thousand dollars. The sale marked a crucial step in solidifying Yale's place in science. More important, it meant that mineralogy, geology, and chemistry were now viewed as serious academic disciplines.

"The object in view presents a powerful appeal to the interest and honour of our city," Timothy Dwight told the citizens of New Haven.

> It is not too much to say, that the Cabinet of Col. Gibbs has, in a great degree, created the science of Mineralogy in this country. It has given our College a high pre-eminence over every other institution of the United States, in the means of instruction in this branch of science, a science so peculiarly important to a new country whose resources are yet to be explored. This Cabinet is viewed with admiration by those who are best acquainted with such collections. It is known throughout our country, and serves, not only to increase the number of those who remain among us for the acquisition of knowledge, but to draw great numbers more to this city, in the indulgence of an enlightened and liberal curiosity.[14]

The sale was notable for other reasons. It pointed to Gibbs's new role as patron of Yale's emerging science program. It was the first time the department had a benefactor. Second, it illustrated Silliman's continuing devotion to education. As a result of Gibbs's largesse, Silliman could now offer an annual prize to a member of the Senior Division of the mineralogical class. The peer-nominated winner could choose a sample from any of the hundred specimens in Silliman's collection, so long as there were duplicates. There was also a Junior Division prize. That winner received a free subscription to the *American Mineralogical Journal* and free admission to Silliman's mineralogy lectures. These were among the nation's very first such awards for academic achievement.

As his reputation grew, Silliman became one of the most popular orators of the antebellum era. He educated and entertained his students using vibrant illustrations and countless minerals and fossils. He conducted amazing chemical experiments to the awe of those who attended his classes, and later he educated audiences on the lecture circuit.

These lectures signaled a new era for Silliman. He became the people's teacher at a time when popular lectures were still something of a novelty. Silliman's public presentations sparked an interest in the study of the physical sciences wherever he traveled. When he spoke in public, it was not unheard of for between three hundred and four hundred people to flock to the lecture hall to hear the tall, dignified, and handsome man with the melodious voice. Once in Boston, the sight of the crowd left him breathless: "The room was more than full, alleys and all. The people filled the stairs, and were clustered around the door in crowds . . . such an audience of intelligent and attentive persons was sufficiently encouraging. The subject of the lecture was meteors. I spoke seventy minutes, giving first an introductory view of luminous meteors. The Weston meteor of December 1807, was fully described, and a summary of the facts was given from my own investigations about the places and among the people where the event occurred."[15]

These sold-out lectures, which guaranteed a retelling of the Weston Fall, appealed equally to the religious and to the secular. Professor Silliman deftly walked the line between science and faith: "Clergymen, both Unitarian and orthodox, thank me warmly. They say that delicate points are treated fairly. They tell me that the success of the lectures is without precedent."[16]

By now Silliman's reputation extended far beyond his laboratory's walls. He began to travel along the East Coast, delivering polished lectures on science. He spoke to a new generation who greeted the material as a kind of revelation. Students looked forward to class, and the public looked forward to his talks. Silliman became a star. In 1839 he came to the new Lowell Institute in Boston to speak about geology. Textile manufacturer John Lowell sponsored the lectures, which attracted many notable speakers. When Silliman came, the resulting stampede for free tickets shattered the ticket office windows.[17]

When the War of 1812 broke out, Professor Silliman and his students at Yale College worried that the British might loot their specimens, for they not only represented the school's material wealth but also America's new-found international academic standing. Quite suddenly, Silliman's dream of an expansive mineral gallery seemed in jeopardy.

"The painful topic was . . . not without an important bearing upon the peaceful pursuits of science. The question of course arose in our minds: Shall we proceed to open more treasures in a maritime town, treasures which we cannot remove and which may be destroyed by the vicissitudes of war? We concluded, however, to trust in god and proceed with our work," Silliman said.[18]

Nevertheless, in spite of the looming peril, the cabinet remained open to visitors. Tourists from near and far flocked to the second-floor gallery to behold the glittering display. To enhance their visit, Silliman printed informative catalogs explaining the collection. The professor's work exceeded expectations.

Not content to limit Yale's collection to the Gibbs cabinet, Silliman pursued other similarly brilliant arrays. A former student of Yale College named Benjamin Douglas Perkins owned an extensive and impressive collection. Perkins was the son of Dr. Elisha Perkins of Plainfield, Connecticut. Like Colonel Gibbs, Perkins had gathered the museum-quality minerals while studying abroad.

Perkins lived in New York City and was eager to sell. This time, based on the success of the Gibbs collection, Silliman easily convinced Yale president Timothy Dwight to buy the Perkins cabinet for a thousand dollars. Silliman traveled to New York to supervise the transfer. The drawers, laden with precious goods, were gingerly removed from their mahogany cabinet and placed on a cart that transported the collection to the dock. A packet ship took the entire cargo to New Haven. The cabinet was then delivered to Silliman's living quarters in the Lyceum, where it remained for several weeks. The cabinet was "arranged in drawers in a case of dark ancient mahogany and when they were displayed successively the appearance was quite captivating." Silliman also described the crystals, jaspers, amethysts, opals, agates, tourmalines, and gold.[19]

News of the splendor spread. Even Jonathan Trumbull, the governor of Connecticut, came to lay eyes on the dazzling shelves. Silliman recalled Trumbull's hesitation to handle the specimens: "[A]nd when I presented to him the beautiful silky amianthus, at the same time handling its delicate threads and offering it to his own fingers, he declined, saying that he would obey the general *noli me tangere* rule of cabinets. I assented, adding, however that the rule was for the many, but as there was only one governor in the State, the precedent could not be followed, and therefore he might handle. The remark was received with his usual courteous smile of acquiescence. I was then twenty-eight years old, and confess I was not a little gratified that the devotion of five years to my profession at home and abroad had been so far successful," Silliman recollected.[20]

Yale now owned two quality collections. Silliman decided the college needed to introduce an independent and elective course in mineralogy. He held the classes in seminar fashion around a table. Minerals were passed on trays like hors d'oeuvres so each student could carefully examine the collection. Silliman used his course to begin rectifying what he considered the problem of scientific education. He understood that merely harvesting facts and naming new discoveries wasn't enough to solidify the country's growing reputation as a center of science. Silliman believed his students needed context and they needed to reconcile science with the Bible. Silliman remained deeply pious but was by no means a fundamentalist.[21]

Just as there were critics of Silliman when he worked on his meteor report, there were some who felt his scientific endeavors were leading the school astray. Silliman recounted the story of how one of the fellows of the Corporation that managed Yale, the Reverend Dr. Ely, asked him, "Why . . . is there not danger that with these physical attractions you will overtop the Latin and the Greek?" Silliman appeased Ely and ensured him that Yale would continue to build its academic reputation.

In 1815, just eight years after the Weston Fall, scarcely any scientific societies existed in the nation. And there were no national scientific journals to merit attention. Here, too, Silliman made his mark. In1818, he published the first issue of *American Journal of Science and Arts.* Still published today, this periodical was the premier general scientific publication for Americans during the antebellum years. The journal, nicknamed "Silliman's Journal," quickly earned widespread respect.

"Although his services as a college officer were great, Professor Silliman's strongest claim to the gratitude of men of science rests upon the establishment, and the maintenance, often under very discouraging circumstances, of the *American Journal of Science,*" wrote the editors of this periodical in an anniversary issue.[22]

Professor Silliman intended the journal to benefit naturalists from all over New England and the nation as a whole. He banked on the idea that contributions from a diverse group of scientists would make the periodical interesting to read, as well as a vital learning tool. Silliman modeled his journal after Dr. Archibald Bruce's *American Mineralogical Journal.* For one brief moment Bruce had been quite optimistic about his publication's prospects, that is, until after the second issue, when it became clear he hadn't enough circulation to sustain it. When the *American Mineralogical Journal* folded, Silliman considered it not only a blow to his friend's ego, but to the national ego as well. More than ever he was motivated to launch his own journal. His relative isolation in New Haven helped protect him from bitter rivalries. Indeed, he enjoyed amicable ties to prominent scientists in several cities.[23]

From the moment of its inception, Silliman's *American Journal of Science* promoted scientific nationalism. During his tenure as proprietor and lone editor for nearly two decades, he published original papers, notices, and reviews. The magazine nourished the country's patriotism and it achieved international accolades. In addition, the journal eventually introduced the newest European discoveries to American readers. And so the domestic audience learned everything from the latest in the natural system of classification, to how to take a chemical approach to mineralogy.

Since its earliest issues, Silliman's journal has enjoyed one of the most prominent positions of any purely scientific journal in the United

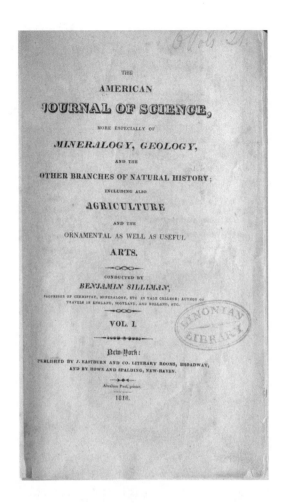

THE

AMERICAN

JOURNAL OF SCIENCE,

MORE ESPECIALLY OF

MINERALOGY, GEOLOGY,

AND THE

OTHER BRANCHES OF NATURAL HISTORY;

INCLUDING ALSO

AGRICULTURE

AND THE

ORNAMENTAL AS WELL AS USEFUL

ARTS.

CONDUCTED BY

BENJAMIN SILLIMAN,

PROFESSOR OF CHEMISTRY, MINERALOGY, ETC IN YALE COLLEGE; AUTHOR OF
TRAVELS IN ENGLAND, SCOTLAND, AND HOLLAND, ETC.

VOL. I.

New-York:

PUBLISHED BY J. EASTBURN AND CO. LITERARY ROOMS, BROADWAY,
AND BY HOWE AND SPALDING, NEW-HAVEN.

Abraham Paul, printer.

1818.

The cover of the
*American Journal of
Science,* founded by
Benjamin Silliman.
Courtesy of Yale
University Manuscripts
and Archives Division.

States. John Griscom, a member of New York City's Literary and Philosophical Society, in a letter to Horace Hayden in Baltimore, remarked that he enjoyed the work and found the journal incredibly informative. Hayden wrote how he was "authorized to proffer cordial support."[24] Jacob Bigelow, Rumford Professor at Harvard College, wrote that Silliman should "[c]onsider me a subscriber to your journal to which I cordially wish success."[25]

Nevertheless, Benjamin Silliman met several obstacles in circulating the journal. The public mail system wasn't reliable, so he hired subscription agents to deliver the periodical. In the beginning, only 400 of 1,200 readers paid their dues, which forced Silliman to finance the publication from private resources. By 1829, a lack of funds threatened to end its publication. In a testament to Silliman's role as the face of science and as the father of science education in America, scores of newspapers across the United States appealed to their readers to contribute to the publication and rescue it from imminent doom.

> Every lover of scientific research must regret the necessity which has again drawn forth the able editor of this valuable work, in an appeal to the public for increased support, and must sincerely deprecate any possibility of its being discontinued for want of subscribers. That a journal which has contributed more than any other to raise our national reputation in the scientific world, should be able to number never exceeding six hundred good subscribers must be considered as far from complimentary to the liberating at least of our countrymen. From motives of patriotism as well as expedience and utility, a discerning public should not allow a work like this to languish. We hope a modest appeal of its editor will not be in vain.[26]

Other like-minded editors asked readers for help. The *Connecticut Mirror* of Hartford asked its readers for help on July 18, 1827:

> It is well known not only in this country, but in Europe as one of the most valuable periodicals of the day; and had probably done more to promote the scientific reputation of this country in Europe, than all the others which have been published here. Indeed it has long ranked with the most scientific journals of Great Britain and the continent, and in the words of the editor has become identified with the progress of Science and Arts of the present day. The reputation and success of this work is not only a matter of gratification to its editor, but also of national pride to us.[27]

In time, new scientists began calling on the professor for advice; he had come a long way from the day when he had traveled to Philadelphia with a candlebox of specimens. Letters from all around the world came addressed to him; from the British Association for the Advancement of Science to the US Legation of Stockholm.

In "Silliman's school," a school in the sense that Silliman had his own set of followers, many young scientists found early guidance and training. His son, Benjamin Jr., followed in his footsteps and later sat on the editorial board of the *American Journal of Science*. Later his son-in-law, James Dwight Dana, edited the journal.

Later his students and colleagues would note Silliman's facility with language. This was no accident. He prided himself on the well-spoken word and often impressed this skill on his students. Language is "one of the most important gifts of God to man. . . . Rhetoric is not therefore, as some imagine, chiefly a collection of tropes and figures . . . it is the accurate, elegant, and energetic exhibition of the thoughts of the mind."[28]

Benjamin Silliman made his mark as a teacher, aside from boosting America's scientific reputation. He attracted both students and the general public.

"In lecturing, his language was simple—his flow of words easy, generous and appropriate. His style animated, abounding in life-like and well adorned description, often eloquent, and sometimes varied with anecdote and running occasionally into wide digressions. His manner was natural, and every feature spoke as well as his mouth; his noble countenance and commanding figure often called forth, as he entered the lecture hall, the involuntary applause of his audience."[29]

Benjamin Silliman devoted his life to seeing that people, whether they occupied the corridors of colleges or the fields on farms, learned a bit of science. He awakened a passion for science and summoned a school of followers across the nation.[30]

He nurtured the sciences more persistently and effectively than any other in the years before the Civil War. He launched institutions and helped establish patterns of patronage to support scientific research. He made science important and accessible. As the first example of a modern scientist in New England, Silliman enabled the American scientific community to come into being.

Chapter Ten

SILLIMAN'S SCHOOL

T hroughout his long academic career, Professor Benjamin Silliman built a reputation that would extend beyond the classroom. His legacy would eventually include a generation of students who made their own impact in a variety of fields and introduced and cultivated his teaching methods in their classrooms. He also offered public lectures, wrote articles, and became involved in pre–Civil War politics.

Most students who passed through the classrooms of Yale College in the mid to latter part of the nineteenth century were fortunate to study at an institution whose name was known not only across the country but also increasingly overseas. Those who chose to pursue science had the good fortune to study under Benjamin Silliman. These first classes were the first generation of students to have official guidance, to benefit from the intellectual exchange that scientific journals and learned societies could offer. Although the field of science remained relatively new in the United States, it had developed considerably. When their professor had entered the field, the discipline was virtually nonexistent. In contrast to Silliman's early years, these students had mentors; they also weren't required to leave home to study abroad, although a great many did venture to Europe for a year or two.

These men, some of whose names are still known, made a lasting impact in their chosen fields of geology, mineralogy, astronomy, and botany. These young men founded schools and chaired departments. Many of them lectured, and all of them published writings, thereby expanding knowledge of science in both the public and the professional arenas. In this way they reflected the aspirations and principles of their teacher.

In the years immediately following the Weston Fall, Silliman was quite naturally consumed with both speaking and writing about the momentous event. He also was spending considerable time editing and publicizing the still new *American Journal of Science*. In short, he also spent the next several decades solidifying Yale College's position as a center of science in America and, in so doing, he established America's early place within the worldwide science community.

The Silliman Family. Courtesy of Yale University Manuscripts and Archives Division.

Although this time period should have been one of immense satisfaction, and in many ways it was, for Silliman it was also marked by several personal tragedies. In 1819, his eldest son, then four years old, died of an illness. Two months later his infant daughter also died of an illness. Then, in 1820 and 1821, Silliman and his wife, Harriet, lost two more infant sons, each just a few short months after their births. Then in 1822, the professor fell seriously ill from what appeared to be a stomach ulcer. By 1823 he had fully recovered and went back to work full time.

Still, Silliman took no respite from work; indeed, he took comfort in the classroom. But he also realized it was time he had some help in the labs. In 1821 Sherlock J. Andrews became Professor Silliman's first teaching assistant. Andrews had graduated from Union College in Schenectady, New York. He helped Silliman conduct all of his chemistry classes until 1823, at which time he decided that perhaps science was not his calling. In a reverse of Silliman's own career shift, Andrews left science to pursue law. At this point, Silliman was fully prepared to appoint his nephew, who was a senior at Yale, as his assistant. But the nephew wasn't interested in science. So Silliman hired Charles Upham Shepard, a graduate of Amherst College in Massachusetts, to assist him in chemistry. Shepard stayed till 1831.

Edward Hitchcock was a young theological student when he first had occasion to hear Benjamin Silliman speak. Silliman's words inspired him, and years later he had the opportunity to work as an assistant in the professor's chemistry laboratory.

Hitchcock was born on May 24, 1793, in Deerfield, Massachusetts. His father, like Silliman's father, was a veteran of the Revolutionary War. The Hitchcock family was poor and could not afford to send Edward to college. Instead, he attended the Deerfield Academy before becoming ordained as a Congregationalist pastor.

While president of Deerfield Academy, Hitchcock decided to introduce himself to the professor he so admired. Together with his letter of introduction, Hitchcock enclosed a box of minerals he wanted named. It reminded Silliman of how as an inexperienced chemist he had brought his valise to Philadelphia in 1803 and asked Adam Seybert for help identifying various pieces of rock and minerals. The gesture touched the professor.

Silliman looked back on Hitchcock's request with fondness: "More than forty years ago, I believe in the year 1817, I received a box of minerals from a person then unknown to me, who signed his name Edward Hitchcock teacher of the Academy of Deerfield Mass," Silliman wrote.

"He stated that he had collected these minerals from the rocks and mountains in that vicinity and as he stated moreover that they were unknown to him, he desired me to name them and return them to him with the labels. I promptly complied with the request and as the accompanying letter of Mr. Hitchcock was written with modest good sense and indicated a love of knowledge I invited him to send me another box, and I promised him to return it with the information he desired. It came and was attended accordingly."[1]

In due course that initial contact blossomed into a professional and personal friendship. Clearly Benjamin Silliman recognized then, in 1817, that Hitchcock had a keen mind most suited for the pursuit of science. While at Deerfield Academy, Hitchcock had been doing his own work mapping the Connecticut River Valley. Impressed with this effort, Silliman asked Hitchcock to submit the geological map in time for the first issue of *American Journal of Science*. "The Journal of Science was instituted the next year 1818 and Mr. Hitchcock appeared in the first volume," Silliman wrote. "His communications have been numerous and important. I have found between fifty and sixty titles of his papers in the tables and contents and in the Index. Not a few of these are elaborate and indicate much care and skill."[2]

Hitchcock contributed to Silliman's *American Journal of Science* from its inception. In 1818 Silliman conferred him with an honorary master of arts degree. The gesture rather surprised Hitchcock, who wrote Silliman a letter on September 28, 1818, to express his profound appreciation.

"The unexpected conferring of a degree upon me by your college awakens within me the liveliest feelings of gratitude and I have much reason for supposing that you, Sir, have not been inactive in my favour," Hitchcock wrote. "Having been completely frustrated in every effort to pass regularly through a college by weakness of sight and of constitution I had relinquished the idea of ever acquiring the honors of any, much less one so eminent as Yale. I fear that I shall never be able to make any adequate return for this favour; but I hope at least that a thankful heart will not fail me."[3] Hitchcock then prepared for the ministry. But, in 1825, he returned to work for a year as an assistant to Silliman. After that he became a professor of science and headed the geological survey of Massachusetts. Later he became the third president of Amherst College.

Hitchcock was one of America's first paleontologists, publishing many papers on fossilized dinosaur tracks in the Connecticut River Valley. He believed that gigantic ancient birds left the tracks, and in some ways he was on to something, since it is now accepted that dinosaurs are closely related to birds.

Deeply religious but highly intellectual, Silliman had successfully married his views on science and religion. His outlook influenced many people, and he certainly guided Hitchcock's view of theology and geology. Hitchcock thus came to believe that one did not have to exclude the other. One of Hitchcock's main interests was natural theology, a field that tried to reconcile science and religion. He published *The Religion of Geology and Its Connected Sciences* in 1851. Hitchcock used the work to argue that the Bible did in fact agree with the newest geological theories. The geologist believed that our planet was in fact far more ancient than the six thousand years so many biblical scholars claimed. One way Hitchcock argued his point was to read the original Hebrew in a way that a single letter in Genesis, *v*, which meant *afterward*, could be interpreted as meaning hundreds of thousands of years.

One of Benjamin Silliman's most superior students and assistants happened to be his son-in-law James Dwight Dana. After Silliman, Dana was probably one of the most renowned early American scientists. When Silliman fully retired in the late 1850s, Dana assumed total responsibility for the *American Journal of Science*. He also was appointed to a position on the Yale faculty. Dana raised money for the Yale Scientific School and lectured extensively on geology and mineralogy. He, too, believed that science and religion could coexist, a point he often spoke about during a long career that lasted well into the 1890s.

Just like his father-in-law, Dana adored exploring different locales to further his education. He would travel to Australia and the Philippines, Hawaii and Fiji. But he would later return to New Haven and Yale, where he studied natural history, as science was sometimes called.

The eldest of ten children, Dana was born in Utica, New York, on

February 12, 1813. He fell in love with natural history at a young age and, in 1830, he decided to matriculate at Yale College. He chose the school largely because of Silliman's already solid reputation and because he had heard about the acquisition of George Gibbs's mineral cabinet.[4] He didn't yet know if he wanted to be a full-time scientist, but he knew he wanted to study geology, mineralogy, and zoology. He was eager to do so under the guidance of Professor Silliman, who was then fifty years old.

Because Dana proved to be such a diligent student, Silliman recommended him as someone who demonstrated both the necessary perseverance and the ingenuity needed to join the Yale faculty. However, before he permanently settled into academia, Dana decided to travel for a year. After he returned to the United States, Silliman offered him a job as assistant in his geological lectures and chemical laboratory.

The job wasn't exactly demanding. It took Dana less than four hours a day to fulfill its requirements, which included displaying rock specimens for the pupils. While it wasn't taxing, it was considered a plum position because in those years very few assistantships existed at major research centers. In fact, there were still very few major research centers in the United States.

Soon the lure of adventure again beckoned Dana, and in 1838 he joined Capt. Charles Wilkes on the United States Exploring Expedition to explore and survey the Pacific Ocean. Dana took what he had learned under Silliman with him on the US navy-led trip. As one of nine scientists aboard the ship, Dana was officially in charge of mineralogy and geology. He also was in charge of cataloging and collecting information about zoophytes and crustaceans. The voyage lasted through 1842 and took him to far-flung ports. While on the trip he studied the Australian coastline, Fiji, Hawaii, and other distant places. Much of what the scientists collected formed the basis of the Smithsonian Institution's early collections.

After he returned to New Haven, Dana married Silliman's daughter Henrietta in June 1844. She was ten years his junior. Soon after Yale appointed Dana as the new Silliman Professor of Natural History, he settled into family and professional life. In 1846 Dana's duties expanded, and he became associate editor of the *American Journal of Science.*

Following his father-in-law's lead, Dana worked on his own books. When he returned from the expedition, he had reams of paper to sort through; he spent the next decade putting all he had seen and studied into three volumes. During that time he also wrote three new editions of his *System of Mineralogy* and a mineralogical textbook. Though the market for textbooks was neither large nor very profitable, it provided Dana a unique opportunity to analyze the field and determine how he could better improve its course of study.[5]

In addition to the textbook, Dana also authored a book titled *Geology*, which was actually a second volume to his original book about the Wilkes Expedition. This work served to confirm Charles Darwin's theory of evolution more fully than Darwin himself thought possible. Certainly all accounts indicate that by 1849, at age of thirty-six, James Dwight Dana had achieved more academic success than many geologists of his day had to show for their entire careers. In 1854 the American Association for the Advancement of Science elected Dana its president.

Like Silliman, he believed in harmony between science and religion. The promising geologist believed that through the study of geology one could better understand the Old Testament, and that his scientific endeavors helped explain the stories of creation and the deluge. Dana had no issue with geologists who interpreted the Bible. He did, however, frown upon those members of the clergy who lectured from the pulpit regarding science, particularly the field of geology.[6]

Over the years Professor Dana assumed a greater role in garnering increased publicity for Yale. He worked alongside Silliman and other former students to help raise funds for various projects. Dana frequently met with alumni about the need to raise money for the school. In 1856, to help raise money for a school dedicated solely to science, he wrote the *Proposed Plan for a Complete Organization of the School of Science, Connected with Yale College.*

Eventually Benjamin Silliman Jr. would make his own unique contributions to American science; however, he would operate under his father's shadow for quite some time. And although Silliman Jr., or Young Ben, as he was often called, never quite attained the celebrity status of his father, he did earn a reputation among his students as an engaging speaker and a dedicated teacher. His father was, of course, most proud of his son, especially when the latter joined his father as his laboratory assistant in 1837.

"But it would be weak and foolish to laud him because he is a part of myself," wrote Professor Benjamin Silliman Sr.

> It would be unjust indeed, for that reason, to withdraw from him his just due. I shall therefore treat him with Roman impartiality and describe him as he is. . . . From his earliest day, however, he manifested mechanical talents of a high order. He was not inclined to folly and he took great delight in visiting shops and manufacatories [*sic*] not only to see but understand their curious processes. In two instances when he was missing from his school for several days together, it was found that in one instance it was to learn the art and mystery of forming the mould for the casting of iron and in another that of making hats.[7]

Born on December 4, 1816, the young Benjamin Silliman Jr. would mature into a most devoted son. His sense of filial obligation manifested itself on more than one occasion. So dedicated was Silliman Jr. that he often subsumed his professional aspirations in order to stay close to his father. This was due in some measure because several of his siblings died in childhood. Silliman Jr. was the fourth child born to his parents and the first son to survive into maturity.

Aside from what he learned at his father's side and during his studies at Yale College, the younger Silliman remained, to a large extent, self-taught in science. He decided to forgo an apprenticeship away from the New Haven campus and he also decided against studying abroad like so many of his peers did. But because of the *American Journal of Science*, which he also helped edit at one point, he did keep up to date with foreign publications. He corresponded with a great many scientists and met many in person.

While attending Yale College, Young Ben joined in campus life and became very involved in the Yale Natural History Society, where he was alternately secretary and treasurer. He graduated with a bachelor of arts degree in 1837.

Young Ben actually discovered he loved natural science while in college. He had always taken pleasure in working with tools. He especially enjoyed spending hours in the shop that his father had built for him in one wing of their New Haven home. The workroom boasted an iron lathe, furnace, anvil, and several other tools.

Timothy Dwight, who served as president of Yale from 1752 to 1817, knew Young Ben from the time he was born until he assumed his post at Yale. The former college president once remarked,

> Professor Silliman, the young . . . had much of his father's geniality. . . . His readiness for conversation fitted him for social intercourse and made his companionship pleasant to his friends. The knowledge and information, which he had gained in different lines, through his studies and his travels, he was ever willing to communicate to others and he was thus disposed to be helpful to them. . . . There was something inspiring for him, no doubt, in this mental attitude, but it was attended at times by a possibility of disappointment. His courage and ardor, however, were unfailing, and he moved forward under their influence toward the end he had in view.[8]

Working as an assistant for his father meant Silliman Jr. learned how to maintain a laboratory, order equipment and supplies, and, of course, how to keep accounts.

"I think it would please you to see the apparatus room now in perfect order and as fast as time will permit I am going through the whole establishment in the same way," he wrote to his brother-in-law Oliver P. Hubbard, who was professor of chemistry, mineralogy, and geology at Dartmouth University.[9] Hubbard also had the opportunity to study under Silliman Sr. for a few years.

In any event, Silliman Jr. grew so comfortable standing before a class of fresh-faced students, that in 1841 he lectured in his father's stead for a full week. Silliman Sr. wrote in his diary: "My son is my right hand and is growing more and more efficient and successful—but by absences in cities (necessary and approved by me) and by illness, he has been away from the laboratory about half the course."[10]

Although he was content in Connecticut, Young Ben periodically considered other professorial positions. In 1849, the Medical Department at University of Louisville, Kentucky, invited him to join their faculty. He moved to Louisville with his small family and taught there for several years. Further following in his father's footsteps, Silliman created a private laboratory for those students who desired to learn practical chemistry. Unfortunately the lab wasn't quite the success he had hoped for. However, a series of lectures he delivered between 1851 and 1852 were exactly the kind to make his father beam. So popular were his talks that hundreds of people had to be turned away. Indeed, his lecturing skills made an impression.

Word of his skills traveled back to the Nutmeg State, and in February 1853, Yale president Theodore Dwight Woolsey wrote to Benjamin Silliman Sr. that he wanted to nominate Young Ben to teach chemistry at Yale College in the still new medical school.

Throughout his time at Yale, Benjamin Silliman Jr. displayed a facility for persuading people to donate rocks, minerals, fossils, and money to the institution. Yet the young Silliman took decided risks in business and pursued certain avenues that at times affected his academic reputation. But his future in industry seemed cultivated by his frequent trips with his father to consult on mines. In 1836 he had the chance to spend two months inspecting Virginia gold mines with his father. While there, Silliman Jr. sat in on geology lectures at the University of Virginia.

In 1855 the son wrote a report about rock oil in Venango County, Pennsylvania. This *Report on the Rock Oil, or Petroleum, from Venango County, Pennsylvania* explained how businessmen use oil commercially. Quite simply his report launched the US oil industry and was used as a reference for fifty years, until the invention of the internal combustion engine. His work in oil also prompted the 1865 Great California Oil Bubble. Silliman had reported about vast quantities of oil floating off the coast of Southern California. Unfortunately the oil was neither easily accessible nor very pure.[11]

Silliman Jr. was the first Yale faculty member to act as a business consultant in addition to teaching his students. He was also the founder and director of the New Haven Gas Light Company and had the first New Haven home with interior gas lighting. In 1847 New Haven became the smallest US city to plan a gas lighting system, and Silliman Jr. helped petition the state legislature for a charter to form the company. Another of his business ventures included being manager of the Bristol Copper Mine in Bristol, Connecticut. His father had originally surveyed the mine in 1839.

Young Ben's interest in mining reached a high point when the Association for the Exhibition of the Industry of All Nations asked him to supervise an exhibit on mining, mineralogy, and chemistry at the New York City Crystal Palace in 1854, which was built for the Exhibition of Industry for All Nations, similar to a World's Fair.

Benjamin Silliman Jr. was a professor, businessman, and fundraiser. He carried on his father's legacy by helping to cultivate Yale's standing, both nationally and internationally, as a center for scientific study. He also helped secure a $150,000 endowment for the establishment of the Peabody Museum from George Peabody. It was the largest single gift ever offered to the school at that time.[12]

Oliver P. Hubbard, who married one of Silliman's daughters, graduated from Yale College in 1828. He had attended Hamilton College in Clinton, New York, until his junior year. On Hubbard's time at Yale, Silliman wrote that "he took a high rank among his classmates, and was greatly respected for his intelligence, his virtues, and attainments . . . and was agreeably remembered by me as an attentive hearer of the lectures, and as indicating by his inquiries both intelligence, curiosity, and habits of observation."[13]

Silliman took Hubbard on in 1828, and he stayed with the professor until 1832. Silliman looked well upon Hubbard and valued his ability to reach out to students: "His punctuality, his exactness in affairs, and perfect integrity, made him entirely reliable, while his knowledge of science in all the branches that belonged to the department qualified him to render efficient assistance."[14]

Charles Upham Shepard was yet another Silliman student to carry on his mentor's tradition to a new generation of students. Shepard taught mineralogy and chemistry at Yale and Amherst Colleges as well as the Medical College of South Carolina.

Shepard, like many of Professor Silliman's other students, often contributed to the *American Journal of Science*. He wrote frequently about meteoric specimens and new mineral species.

"He manifested the deepest interest in favoring and assisting me in all my studies, permitting me to examine freely the treasures of the Gibbs Cabinet of Minerals, then the only one of note in the country," Shepard wrote to a Silliman biographer in August 1865. "[He] encouraged me to engage in chemical researches, accompanied by the generous permission of ordering for the Laboratory whatever might be needed in their prosecution."[15]

Shepard first heard about Silliman during his boyhood in Rhode Island. He read with relish the professor's account of his 1805 trip to Europe. Impressed by the book, Shepard began collecting minerals and started corresponding with the professor. It would be nearly twenty years before Shepard would actually meet Silliman. When he was finally introduced to his academic idol, Shepard impressed the professor.

"His manners were amiable and gentlemanly and moral character pure," Silliman wrote. "Mr. Shepard was already proficient in Mineralogy, and his services were at this time particularly acceptable in that department as I was now to resume the lectures in the cabinet which had been suspended or imperfectly given of late. He was also, to a considerable extent, acquainted with geology, and was advancing in both of these departments. He had formed habits of traveling to observe localities of minerals, fossils &c and his views were directed to science as the business of life."[16]

As was the case with many of Silliman's students, Shepard's life intertwined personally and professionally with the professor. He asked Harriet Taylor, Silliman's adoptive daughter, to marry him in 1831. Shepard helped Silliman in the laboratory for five years before striking out on his own.

Time and again throughout his career he would help his mentor. In 1832 the government asked Benjamin Silliman to look into the sugar cane industry to determine more efficient processing methods. As a result,

Shepard and Oliver P. Hubbard helped him survey several southern states. The resulting paper, "Investigations of the Culture of the Sugar Cane and Manufacture of Sugar Crude and Refined," was published.

Shepard paid homage to his teacher in the dedication of part 1 of his 1832 *Treatise on Mineralogy*: "The distinguished services you rendered to the cause of Mineralogy and Geology in America, not merely in contributing to their early introduction, but to their cultivation among us. . . . The wisdom and zeal of those exertions which secured to Yale College the most splendid collection of minerals in the country, the valuable instruction and enthusiasm imparted by your lectures to a great body of young men now dispersed through the whole nation."[17]

Shepard shared one of Silliman's great passions—meteorites. He began collecting them in earnest, gathering as many fragments as possible in order to learn from where they came and from what they were made. He observed the sky and collected information about shooting stars and fireballs, and he filled many notebooks with folk stories about meteorites. His journals recorded how meteorites were found and where the samples ended up. In this way he carried on Silliman's initial work by writing in the *American Journal of Science* about various meteorites and fragments, and their composition.

Many students like Shepard carried on Silliman's legacy by instituting his teachings and practices in their own classrooms, from taking students on field trips to having them handle specimens.

"As I look back to 20 or 25 years ago, it occurs to me as a curious reflection that four persons who were members of your class, or in your laboratory have now under their instruction by lectures, some hundreds of young men in the same or very similar branches—I refer to dr. Hitchcock, Prof. Hubbard, your son, & myself. Thus may the useful work go on," Shepard wrote to Silliman in 1851.[18]

Aside from the many students he mentored, Professor Benjamin Silliman Sr. left his mark on several institutions, including the Yale School of Applied Chemistry.

From the time he investigated the Weston Fall to the time he fully retired, Silliman never ceased worrying about and working toward ensuring that Yale would become a true center for scientific greatness. Yet, for so many years those studying science did not have their own schools or curricula to follow. Finally, in 1846, working together with his son and John Pitkin Norton, a professor of agricultural chemistry and of vegetable and animal physiology, the professor established Yale's first scientific school. It became known as the Yale Scientific School in 1854 and ultimately the Sheffield Scientific School in 1861. Graduate-level courses were not offered until 1861.

As part of the growth of the still new course of study, the Yale College Corporation approved two new professorships in applied science in August 1846. It was, of course, Benjamin Silliman Sr. who convinced the ever-conservative administrative body to create these two new posts. One of the positions went to John Pitkin Norton. The other opening went to Benjamin Silliman Jr., who was named professor of practical chemistry.

Upon assuming their new positions, the two professors took over the former residence of Jeremiah Day, who had been president of Yale from 1817 till 1846. They remodeled the white clapboard home into a functioning laboratory. Not too long after they started, the young professors realized they lacked adequate funds for upkeep. The college tightly gripped its purse strings, compelling the pair to draw money from their personal accounts to pay for new equipment. This was in spite of the fact that the School of Applied Chemistry, as it was then called, was officially part of the Department of Philosophy and the Arts.

The Corporation maintained that it was really no more than a loose affiliation: "The support of this professor is in no case chargeable to the existing funds or revenues of the college."[19] In fact, in the beginning students of the chemistry school didn't need to meet entrance requirements; they just had to show an interest in practical scientific study. The school started nevertheless with nearly no funds.

Over time Silliman Jr. and Norton tired of this arrangement. They changed the structure of the classes and petitioned the school to allow students to earn a degree in applied chemistry. Finally, in 1852, the Yale Corporation agreed to formalize the school. Students who had completed course work and examinations could earn a BA in philosophy.

The new school taught students how to apply chemistry to practical matters such as manufacturing. There was even a course for farmers who wanted to learn basic principles of science agriculture.

The School of Applied Chemistry didn't stand alone for long. In 1854 it was combined with the School of Engineering to form the Yale Scientific School. As part of their mission to develop the chemistry curriculum on campus, Silliman Jr. and Norton founded the Berzelius Society in 1848. This organization, named for the Swedish chemist Jöns Jakob Berzelius, was intended as a scientific society for students to discuss chemistry and, from time to time, to hear a speaker from the field. Here, Silliman Sr.'s legacy was felt as the school grew in stature. He wanted to give a course of lectures to raise money for the society in 1852 but was unable to due to ill health. In time the society became a secret society, similar to Skull and Bones.

Professor Benjamin Silliman Sr. influenced his students in areas outside of academia as well. For one, he taught them the importance of cultivating benefactors. In 1858 Silliman Sr. convinced New Haven's Joseph Earle Sheffield to donate property for the school in addition to a large endowment. Sheffield, a wealthy financier, purchased the former Medical Department on College Street and supervised the building of two additional wings. He had the inside completely remodeled and then gave the entire building to Yale. He also donated $50,000 to the school to endow chairs in engineering, metallurgy, and chemistry.

In 1860 Professor Silliman thanked Sheffield: "I should not have presumed so much on your kindness had I not been gratified, more than I can express, by the happy progress of our truly good and useful institution under your auspices. As the original mover of this enterprize [*sic*]—as its Sponsor on the Baptismal Font, I may be pardoned, I trust for my freedom."[20]

Nearly fifty years after Silliman had taken his first tentative steps in the field of chemistry, he realized a school dedicated to carrying his vision forward.

In addition to his constant work as America's public face of science, and his endless efforts to raise money, Silliman Sr. never stopped stepping out to speak about science. The first of these lectures came in 1834. Residents of Hartford, Connecticut, pressed him to give a series of lectures on geology. He obliged them in the spring. Based on the success of that lecture he was invited to Lowell, Massachusetts. Then in 1835, the Boston Society for Promoting Useful Knowledge invited him. He spoke in the Masonic Temple in Boston before 1,200 people.

Benjamin Silliman also began traveling more in his later years. He returned to Europe for a second time in 1851. He was seventy-one years old. It had been forty-six years since his last trip, and he spent time visiting scientific societies and calling on colleagues. He went from England to France, Italy, Sicily, Switzerland, Germany, and Belgium. He wrote and published a two-volume detailed book about this trip. And two days after his return home he married his second wife, Sarah Isabella Webb, on September 17, 1851.

In 1853, at seventy-four, he retired from teaching, at least to some degree. He stayed on at the Yale faculty as professor emeritus and continued delivering lectures in both mineralogy and geology. During his retirement Silliman Sr. took the time to gather together his copious amount of correspondence, papers, and diaries. He wrote and promoted Yale. He cultivated supporters and he dipped his toe in the most contentious political and moral debate of the time—slavery.

The Silliman family had owned slaves; his mother, with eighty-five slaves, had owned the most slaves in Fairfield County. Through the years the professor became a staunch abolitionist. His beliefs were sharpened during his travels throughout the country as he researched, surveyed, and lectured. Once in Virginia, while working on his 1836 report on sugar cane, he wrote about seeing slaves. This entry is notable for its shift in attitude from what appears to be condescension to fierce outrage:

> In our travels, and in the prosecution of our researches among the mines, we had of course met with slaves everywhere, and in general they were, as far as we observed, treated kindly. . . . As we were quietly reading in our apartment at a tavern, on the Sabbath, the landlord entered, with an apology for the intrusion, and, opening a glass case in

the corner of the room, took down a large riding coach-whip, which we supposed was wanted from some excursion. Within a few minutes, however, we heard through our open windows the sharp reverberations of the lash rapidly repeated, and accompanied by loud cries of distress. . . . Slavery, begun wrong, is sustained by a cruel despotism.[21]

In 1854, eighteen years after Silliman commented on slavery in Virginia, Stephen A. Douglas, a senator from Illinois, pushed through the Kansas-Nebraska Act in Congress. The act made it clear slavery was going to be allowed to spread west—a place from which it had been barred. Yale's student body simmered; students started to separate along state lines, and in a few years the Southern students would withdraw from the college.

Benjamin Silliman denounced the Kansas-Nebraska Act in a public meeting in New Haven. He also helped organize Yale's giving of twenty-seven Sharps rifles to soldiers heading to Kansas. The US Congress organized committees to discover who had been arming soldiers with these rifles.

Silliman began his career on the cusp of a scientific awakening. His lifelong effort to ensure science teaching and research in the United States was realized. His influence was so far-reaching that his students helped expand the reach of science, both within the confines of Yale and across the country. The scientists he trained had a profound influence on the world's understanding of science. Professor Silliman never stopped guiding these men until his death in 1864.

Chapter Eleven

WHERE WE ARE TODAY

It's nighttime, April 10, 2010. The sky over Livingston, Wisconsin, suddenly turns a greenish hue. Streaks of light shoot across the sky for nearly fifteen minutes, a light show that lasted far longer than the one during that winter morning in 1807 in Weston, Connecticut. Many people were shocked, and some were scared, just as they were nearly two hundred years ago. Calls to police, first responders, and even the news media immediately ensued.

Meteorites still have the power to awe, no matter how much science has taught us. It still seems that an event such as the Weston Fall or the most recent fall over the Midwest can trigger something primeval, awakening a part of ourselves long considered dormant. But in truth the fear some people have regarding unpredictable weather and related phenomena continues to live alongside other antiquated suspicions and superstitions. Some people react similarly to those who saw the Nogata meteorite fall in Japan in 861 CE and to those who witnessed the meteorite fall on Ensisheim, Germany, in 1492. Every time nature stages a dramatic show like the Weston Fall, it gets our adrenaline pumping and grips our fascination.

The bright light that flamed across the midwestern American sky on April 10, 2010, was—of course—a meteor. In the aftermath, it was learned that people saw the fireball over Wisconsin, Iowa, Illinois, and Missouri around 10:15 p.m., local time. Just like the Weston Fall, this meteor rendered the night into day, and the force unleashed a sonic boom that reverberated for hundreds of miles. The thunder-like rumble was especially impressive near Livingston, Wisconsin, a town that straddles Iowa and Grant counties about twenty miles southwest of Dodgeville.

Bob Dowr, a bartender at the Alley Oops Tavern in Livingston, told reporters that "[e]verybody has been talking about it since it happened. It basically got light as day here. And it rumbled like a trolley car."[1]

However, unlike the Weston Fall, film captured this fireball as it blazed across the sky. Dashboard cameras in police cars from Wisconsin to Iowa filmed the meteorite as it passed overhead. Amateur videographers caught the sight on cell phones and small recorders.

The evidence indicates that the fireball might have been nearly six feet wide and, before it broke into dozens, if not hundreds, of pieces, it might have weighed about one thousand pounds.[2] After all, if an object is bright enough to make the night sky appear like daytime and send sonic booms rippling for hundreds of miles, it has to be large. Each fragment that fell was probably about the size of a football or smaller. If the meteorite had fallen during the day, astronomers said people would certainly have seen smoke trails following the bright light.

The timing of the Livingston Fall happened to coincide with the yearly Lyrid meteor shower, which occurred at the same time. Meteorites actually have no connection with meteor showers, which usually result when Earth passes through the dusty tail of a comet. Meteorites can fall from the sky at any time.

Thanks to the work of Professor Benjamin Silliman, people now know that meteorites come from outer space. Even so, teams of scientists, meteorite hunters, amateur collectors, and astronomers all converged on the midwestern fields, and they continued do so for months, because the study of meteorites, and their parent bodies, remains important. In this way the spirit of Benjamin Silliman, Isaac Bronson, James Kingsley, and others involved in the Weston Fall still influences modern meteoritics.

Today, when scientists study meteorites they hope to learn at least five points that relate not only to the field of meteoritics but to other scientific disciplines as well. First, scientists hope to glean more information about the evolution of stars. Some meteorites contain both chemical specimens and grains of dust that stars produced before the solar system was born. When scientists study these samples they learn even more about stars, their composition, and the unique role they play in the universe.

Second, meteorites can teach us about the evolution of the solar system itself. Contained within the different components of these rocky chunks is information about changes in the conditions of the chemistry, temperature, and pressure of the early solar system.

Third, when scientists study and measure specific ingredients of meteorites, they are more able to estimate both the age of our solar system and how it was formed.

Fourth, examining these specimens from space allows scientists an opportunity to deduce the geological history of both Earth and the moon. Large meteorite impacts, such as those at Vredefort, South Africa; Popigai, Siberia, Russia; Chicxulub, Yucatan, Mexico; and Chesapeake Bay, Virginia, in the United States, shaped the topography of both Earth and its satellite moon. These events are also thought to have altered Earth's climate. It is accepted scientific theory that a meteorite prompted the extinction of the prehistoric dinosaurs and ushered in the Ice Age.

Fifth, and last, scientists study meteorites to learn the history of life. There is much evidence to suggest that the necessary chemicals for the origin of life on Earth might have arrived on our planet by meteorites.

Professor Benjamin Silliman's work continues to serve as a foundation for modern scientists. His influence can still be felt in current research on meteors.

In 1807, competition among scientists was fierce. They wanted to prove their worth to potential sponsors, but mostly they wanted recognition for their work. Silliman had to contend with rivals such as Nathaniel Bowditch and James Woodhouse, both of whom tried to dispute his work. He had to race against the interests of farmers eager to sell and profit from pieces of the Weston Fall rather than give them up for scientific analysis. Today, the competition among those engaged in science is considerably more intense. Because of e-mail, Twitter, and other technological communications breakthroughs, word of the Livingston Fall spread quickly in real time. Scores of meteorite hunters headed for southwestern Wisconsin in search of treasure from outer space. One hunter, Ruben Garcia, known as "Mr. Meteorite," travels with seven colleagues and a meteorite-hunting dog named Hopper.[3]

As these meteorite hunters and scientists collect fragments from the Wisconsin event, they will be using data, formulas, and some experiments that got their start in a New Haven laboratory more than two centuries ago. These fragments, like those at the Weston Fall, will add valuable knowledge to the field of meteoritics. These newest additions, older than any rocks on Earth, will help propel the study of everything from Earth's origin in the solar system to the length of time Earth has been in outer space.

Today, such specimens are prized for both their beauty and their composition. They can sell for more than five hundred dollars a gram, which is barely 1/28th of an ounce. Of the meteorites from Mars, which landed in Antarctica nearly four billion years ago, only twelve specimens have been recovered. They sell for more than one thousand dollars a gram.[4]

Between the mid to late 1990s there was an increase in the number of meteorite collectors competing for the limited amounts of meteoritic material appearing on the commercial market. Now serious collectors often visit Internet sites as well as mineral and gem shows in search of exceptional meteorite collections from reputable dealers. Some specimens will end up on the auction block, some behind glass cases in museums, and others will adorn mantels in homes for years to come.

Once again meteorite mania has gripped the nation. In 1996 the Pathfinder Mars probe ignited renewed interest when evidence of primitive life was discovered in a Martian meteorite. Currently it is the Livingston event that has people again looking to the skies and scouring the ground.

Nearly three decades after Silliman investigated the Weston Fall and delivered his paper both in America and abroad, the world of meteoritics had another discovery. In 1833, Denison Olmstead, a physicist and astronomer from East Hartford, discovered that meteor showers originated from a common point in the sky. Olmstead realized that meteors seem to diverge from a point in the sky because they originate in a swarm of meteoroids moving on parallel paths through space. In 1866 the Italian astronomer and science historian Giovanni Schiaparelli discovered that the comet "1862 III" was causing the Perseid meteor showers, which occur in August. And the discoveries continued.

Since those years, the links between the field of meteoritics and planetary astronomy, astrophysics, analytical chemistry, and the study of our origins have strengthened immensely. The 1807 Weston Fall is still considered a definitive moment in American science history. The Weston meteorite itself is extremely famous, and its history and composition is still taught in planetary science. Scientists now understand that together the composition, chemistry, and mineralogy of meteorites yields evidence for a wide variety of chemical and physical processes, including a crucial understanding of how solar systems in general came to be, and, of course, about ours in particular. Clearly, the field of meteoritics has made great strides since Silliman published his report before the Connecticut Academy of Science in 1809. Scientists now understand that meteoritics play a large part in understanding the astrophysical processes of star and planet formation.[5]

Today, astronomers look into the future by looking into the past. Only recently have scientists begun unlocking the historical records inside these chunks of rock and metal. What they discover is most revealing.

Benjamin Silliman has the distinction of being the first American scientist to analyze fragments in a laboratory, and his efforts resulted in some rather accurate measurements. But now, analyzing the fragments is a highly precise operation. Since he separated the Weston Fall's components, scientists have been able to isolate diamonds, silicon carbide, graphite (a form of carbon), corundum (aluminum oxide), and silicon nitride. Scientists now have tools such as electron microscopes, which let them see minute traces of other minerals such as carbides of titanium, zirconium, and molybdenum.[6]

Moreover, scientists can actually see meteors before they become meteorites. The Hubble Space Telescope can relay images of planets, stars, and other gaseous nebulae back to Earth, providing tantalizing clues about how Earth and other planets were formed.[7]

The ability to peer into space before that vast area beyond sends its rocky messengers to Earth is a tremendous achievement. Having this capability furthers the chances for understanding how exactly planets like ours came to be by observing more about the activity in space that preceded their formation. And scientists know what ingredients to look for, water being a vital one. As the search for life or the ability to host life on other planets continues, it is known that some form of water must be present for life as we know it to begin. Meteorites actually offer key clues about how Earth got its water. During the bombardment phase of Earth's formation, any water that meteorites contained would have been incorporated into their crust.

Meteorites are also thought to carry the biogenic molecules that caused life. This concept, that the seeds of life came from meteorites, has a long history. Benjamin Silliman and James Kingsley touched upon this with their study "An Account of the Meteor." Scientists have built upon this theory ever since.

In the nineteenth century, German chemist Otto Hahn wrote about various structures inside iron meteorites that he conjectured were petrified remains of algae and ferns, and fossilized corals, sponges, and crinoids in chondrites.[8] Hahn received a torrent of criticism for his position, but he persisted. Then in the 1930s, bacteriologist Charles Lipman cultured living cells from meteorites. His work was met with scathing reviews, and for good reason.[9] No one at that time had figured out that lab equipment used to study inanimate objects must be as sterile as medical equipment. However, the idea to sterilize equipment took a long time to take hold. Even as late as the 1960s, a scientist named Bartholomew Nagy found small bits of matter in meteorites. At first Nagy and his colleagues announced they had discovered fossils. Actually, they had discovered little more than starch and ragweed mixed with mineral grains.[10]

Naturally these missteps meant that the scientific community, not to mention the public, remained most skeptical and suspicious regarding the notion that meteorites helped usher life to Earth.[11] But then, in 1996 David McKay and others from NASA's Johnson Space Center and Stanford University proposed that a Martian meteorite, ALH84001, recovered in Allan Hills, Antarctica, contained biochemical markers, biogenic minerals, and microfossils of extraterrestrial origin. That meteorite, which a team of US meteorite hunters found, crystallized about 4.5 billion years ago and is older than any other known Martian meteorite. This discovery has opened up the idea that perhaps planets trade rocks; that life was somehow transported from one planet to another.

Aside from studying meteorites to determine Earth's tumultuous past, most scientists agree that it's imperative to study meteorites in order to predict, and try to prevent, a future catastrophe.

Potentially destructive meteorites, similar to the one that wiped out the dinosaurs, explode about once a year. These meteorites are similar in power to the Hiroshima nuclear bomb detonated in 1945. Unlike the Chicxulub meteorite in Mexico, these meteorites usually occur high in the atmosphere and don't damage Earth. There are, however, many smaller asteroids roaming the outer reaches of space. If they were to impact our planet, they would create a crater several hundreds of miles across. Luckily, these kinds of impacts usually occur only about once every ten thousand years. Moreover, as devastating as these meteorites could be, their destruction would be somewhat localized. Truly global catastrophic impacts occur over extremely long periods of time. Iron meteorites create most craters the size of Chicxulub, but meteorites with large iron deposits are far less common than stony meteorites. Furthermore, Earth has natural defense mechanisms. First and foremost, our atmosphere generally causes meteorites to break apart as they enter.

Still, a considerable amount of interplanetary and interstellar matter falls down on Earth each year. With each new bit of spatial dust, much more knowledge is added to the story of how Earth came to occupy its position in the surrounding interstellar neighborhood.

Today, planetary scientists continue searching for clues to unravel the past events. They develop hypotheses and construct theories, many of which remain controversial, a fact that hasn't changed much since Silliman's time. Modern scientists still rely on observation, measurement, and experiment, but they also enjoy the luxury of better equipment. Since the nineteenth century, the field of meteoritics has greatly developed to include advanced physics, chemistry, geology, and astronomy.

Since Silliman's time scientists have come to consider asteroids much more carefully. After all, a meteorite was likely responsible for the extinction of vast numbers of species along with the dinosaurs and quite possibly brought about the Ice Age.[12] But even with all the groundwork Silliman laid, it wasn't until the 1930s and 1940s that scientists truly began to accept that meteorite impacts may well have caused the multitude of craters found on Earth. Yet the knowledge that meteorites caused actual earth-shaking events didn't really come into play until the late 1960s. To be sure, it wasn't until the 1980s that scientists discovered chemical evidence that suggested a huge chunk of rocky detritus hit Earth sixty-five million years ago at end of the Cretaceous period, causing the extinction of 70 percent of animal and plant life on the planet.

Thus meteoritics is no longer viewed as the fringe field it once was in Silliman's day. In fact, it is just the opposite; scientists reach to the stars through their telescopic lenses rather than wait for bits and pieces of the universe to fall to Earth.

In spite of the number of meteorites that fall to Earth or pass us by every year, most are never recovered. On average, only between five and six meteorites a year are actually witnessed falling and then subsequently recovered. That was certainly the case for the December 1807 Weston Fall and the April 2010 Livingston Fall. For the most part, people simply find meteorites without ever seeing them plunge from space.

Incredibly, despite the number of meteorites noted in recorded history, there is no evidence that one has ever struck and killed anyone. Still, there have been close encounters. For example, in 1984, two friends,

Kathleen Clifton and Theresa Davies, were vacationing on Western Australia's Binningup Beach. Suddenly a whistling noise and a thud cracked through the morning sounds of waves lapping ashore. Clifton called to her husband, who, for just a moment, thought a sniper was attacking the sunbathers. Instead of a bullet in the sand, he found that a black stone had landed about thirteen feet from his wife's beach towel. The stone was cool; their nerves, not so much.[13]

As is the case with most falls of that magnitude, the fireball, which can last mere seconds, was witnessed streaking across the sky miles away from where bits of it fell. Residents in nearby Perth heard two loud thunder-like bangs. Though searchers left no stone unturned, no more fragments were found.

The more fragments scientists can gather, the more complete will be the picture of the event. So meteorite hunters are helpful. That's why Dr. Isaac Bronson acted as an intermediary after Professor Silliman went back to Yale. That's why one of the first meteorite hunters in the United States got started. Harvey Harlow Nininger was a biology teacher who became interested in meteorites in the 1920s while traveling through the Midwest. Along the way he started teaching people how to recognize meteorites and, in doing so, he accumulated many specimens.[14] Silliman paved the way for meteorite hunters and the valuable service they perform.

The knowledge of just what lies hidden inside a meteorite began to expand in 1807 when Benjamin Silliman cracked open a fragment in his small Yale laboratory to explore its contents. Since then information about their mineralogical deposits found in meteorites has increased tenfold. A parade of scientists has followed Silliman's lead.

In the 1920s, one scientist, V. M. Goldschmidt, made it part of his life's work to learn the differences in elemental composition between meteorites and Earth rocks. He discovered that their chemical elements had many similarities. He learned that metallic elements, like gold, are usually found as metal alone or alloyed with other metals. Other elements inside rocks or meteorites included silicon, magnesium, and calcium, which form oxides and silicates. Finally, there were gaseous elements, such as xenon, neon, and nitrogen.[15] Goldschmidt's work built on Silliman's initial report, which named and measured the components of the Weston Fall in great detail. Using this intricate analysis, Goldschmidt was able to explain the principal chemical components of Earth, thus showing that Earth rocks and meteorites from space shared many chemical elements and may therefore have originated from common sources in the universe. Subsequently, Silliman's original nineteenth-century study spawned a new branch of geology called geochemistry. That in turn paved the way for modern planetary science.

Sir Arthur Conan Doyle, the creator of Sherlock Holmes, once wrote: "When you have excluded the impossible, whatever remains, however improbable, must be the truth." From Benjamin Silliman to the meteoricists of today, scientists practice the same principle as the imaginary detective from Baker Street. Of course, sometimes the only evidence scientists have is the fragment itself. For although it's known that meteorites are cosmic debris, it's not known exactly how representative meteorites are of early solar system materials. In the beginning, it seemed that most meteorites were ordinary chondrites, or stony meteorites. But then scientists began to explore the icy and dry continent of Antarctica and other regions of the globe. More of the world's deserts were searched. These expeditions yielded troves of meteorites, and it became clear there were other kinds of meteorites. As the field continues to grow, it is clear that most meteorites come from a large number of small asteroids, all very different in composition.

Knowing that there are different kinds of meteorites, something Silliman would not have been aware of, helps us to more fully understand the big bang theory of the origin of the universe—a theorized cosmic explosion that enabled the formation of the universe.[16] Every time a meteorite falls, it sparks more interest in science, astronomy, physics, and geology.

In the past few decades, that interest has led to many more space missions to fill gaps in our existing knowledge. Scientists have become proactive in their search for stardust in space. They know that like meteorites, most of a comet's components contain isotopic compositions (chemical structures) similar to Earth's and originate in our solar system.

In 1999 NASA launched just such a mission, named STARDUST. A probe was launched on a seven-year mission to fly past the nucleus of comet Wild 2 in order to collect particles. The probe reentered Earth's atmosphere in January 2006. When it returned, scientists were not surprised to find spectacular silicate crystals in the comet, something already indicated by astronomical observations. However, the mission results provided other important new insights into a comet's origin and history.[17]

In the sample of comet Wild 2, NASA scientists discovered it contained glycine, a fundamental building block of life. Living organisms use glycine, an amino acid, to manufacture proteins. This was the first time such an amino acid had been found in a comet. The discovery supported the theory that some of life's ingredients formed in space and were delivered to Earth long ago by meteorite and comet impacts.[18]

Another trip to space in 2003 built upon the STARDUST discovery. Europe's Space Agency launched ROSETTA in January 2003 to intercept the comet 46P/Wirtanen in November 2011. Named for the Rosetta Stone, which was found near Alexandria, Egypt, in 1799, this mission will also pass near two asteroids, Otawara and Siwa. ROSETTA will never return to Earth, but it was outfitted with onboard equipment designed to relay images and data to scientists. The information will tell them about the composition and makeup of small solar system bodies. Like Silliman, who took apart and analyzed meteorite fragments, ROSETTA will also help decode the origins and ingredients of meteorites.

Silliman's groundbreaking research not only influenced science, in some ways it also indirectly influenced popular culture. People have long been fascinated with doomsday scenarios since the advent of mass entertainment. Many of these fantasies involve space invaders and asteroids.

Recently the much derided film *2012* featured earthquakes, meteor showers, and even a tsunami that dumped an aircraft carrier atop the White House. In the 1998 film *Armageddon*, NASA sends a coterie of blue-collar deep-core drillers to deflect an asteroid on a collision course with Earth. In the 1998 film *Deep Impact*, it's a comet that is about to fall to Earth and destroy the world as we know it. Astronauts are charged with bringing a missile up to space and shooting down the potential annihilator. As far-fetched as these scenarios are, today's scientists don't want to be caught by surprise. Asteroids, comets, and other deep-space objects are constantly monitored.

For example, in 2004, scientists at NASA's Jet Propulsion Laboratory in Pasadena, California, noted that a nearly 1,300-foot-wide asteroid was hurtling toward Earth, destined for Europe, India, or Southeast Asia.[19] They forecast it would arrive in the year 2029 and had a 2.7 percent chance, or one out of thirty-seven, of hitting Earth.[20] Not wanting to gamble, the scientists repeatedly examined its trajectory. They eventually learned that it has a zero chance of collision. Still, in a once-in-every-800-year event, the body will pass closer to Earth than any current communications satellite in space.[21] As it approaches Earth, some 18,300 miles above, it will likely appear as a moderately bright point of light moving quickly over the mid-Atlantic Ocean sky.[22]

The asteroid, aptly named Apophis after the Egyptian god Apep who lived in eternal darkness and sowed destruction, is slated to return in 2036, with a 1 in 45,000 chance of hitting somewhere in the Pacific Ocean, in a zone somewhere between California and Central America.

Because Apophis is so much smaller than Earth, its orbit will be affected as it passes by our planet. Astronomers found that the asteroid's orbit will actually bend about twenty-eight degrees, which means Apophis will not only have a larger orbit; it will also travel more slowly. And for people still worried that Apophis means annihilation, they can be comforted by the fact that they have a greater chance of being struck by lightning. Apophis won't wipe out humankind, but if it did slam into Earth, it could create a tsunami exponentially far more powerful than the 2004 Indonesian tsunami that killed more than two hundred thousand people.[23]

Scientists now know, unlike during Silliman's time, that there are literally millions of asteroids hurtling through space at speeds of 28,000 miles or greater per hour. These asteroids range in size from 560-mile-wide Ceres to those no bigger than a large SUV.[24] They travel in the asteroid belt between Mars and Jupiter. Of course, Jupiter's gravitational force prohibits the matter from ever fusing together to form a planet. If it did combine, it would have a mass smaller than Earth's moon. Some scientists think Phobos and Deimos, Mars's two moons, might actually be captured asteroids. But in our solar system, Jupiter seems to be the planet to toy most with the orbits of asteroids. Some are ostracized from the asteroid belt and sent hurtling on their way toward the sun, deep space, or beyond Pluto.

Nevertheless, most scientists agree that it's not a matter of if an asteroid will hit Earth but when. Benjamin Silliman shared this sentiment. The Yale professor often asked his audiences to consider that a meteorite might one day devastate entire cities and mountains, causing unforeseen catastrophe.[25] In 1908, a meteorite hit the Tunguska region of Siberia and leveled a forest. Fifty thousand years ago, a three-hundred-foot-wide meteorite plunged into the Arizona desert, leaving a three-mile-wide crater.

Just as Silliman encountered skeptics regarding the origin of meteorites, modern scientists encounter raised eyebrows when they speak of asteroids crashing into Earth. For so long, scientists and astronomers have not taken the threat of asteroids seriously, instead saying that they harmlessly float through space between Mars and Jupiter.

However, as researchers began to take note of impact crater sites across the world, the idea that perhaps asteroids have altered evolution and climate has astronomers taking a more serious look. Today, NASA continues to investigate ever-more precise ways of locating and tracking asteroids with its Near Earth Object Program, which detects, tracks, and characterizes potentially hazardous asteroids and comets. The program aims to locate as many as 90 percent of the estimated thousand asteroids and comets that approach Earth. So far, more than seven thousand near-Earth objects have been discovered since the program started.[26]

According to many astronomers, physicists, and engineers who study and track these masses, their work will greatly benefit the world. Many scientists believe that asteroids contain valuable raw materials that could help in the engineering of space structures and generating the rocket fuel needed to explore and possibly colonize the solar system. Analyzing these ingredients has even led some people involved in space exploration to conclude that the asteroid belt between Mars and Jupiter contains mineral wealth equaling an estimated $100 billion for every person on Earth today.[27] In addition to asteroids, comets, too, contain essential components. These objects are loaded with water and carbon-based molecules needed to sustain life. Water from comets yields liquid hydrogen and oxygen, two necessary ingredients for rocket fuel.

When the Weston meteorite crashed into farmers' fields in 1807, there were many who looked upon the stony fragments as keepers of great wealth. And although men like William Prince and his sons did not find, and would not find, gold and silver within the meteorite, they correctly concluded that space contains treasures. It took a man like Professor Benjamin Silliman to lay the groundwork for what scientists know today, and it was his work that has gifted modern scientists with the imagination to continue to unlock the rich mysteries of space.

EPILOGUE

After returning to his childhood home in 1864, Westonite Edmund Turney published the poem "The Meteor."

> T'was here that the meteor broke,
>   And scattered its fragments athwart;
> Here fell the aerial rock,
>   In the plat? Of mine infancy's sport.
>
> For a moment it spread o'er the night
>   The dazzling effulgence of noon;
> It had suddenly gleamed on the sight;
>   It faded and vanished as moon.

And then two later stanzas:

> And how oft did the query arise,
>   Whence, whence could this visitant come?
> Or why should this Child of the skies
>   Thus seek on our planet a home?
>
> What laws had its action controlled
>   Where none its dark pathway could trace?
> Or how long, all unseen, had it rolled
>   Through the depths of ethereal space?[1]

Following the poem's publication, Turney began to correspond with Professor Silliman. Their letter exchange illuminated Silliman's continued fascination and nostalgia for the day the meteorite fell.

"Your reminiscences of the Weston meteor, revive that subject in my mind very vividly, although more than fifty years have passed since in company with my late friend, Prof. Kingsley, I explored the facts at the places where they occurred," Silliman wrote. "It was a magnificent phenomenon, and remains to this day among the most remarkable occurrences of the kind that are on record."[2]

A magnificent phenomenon indeed, it gave birth to both the career of the young scientist and science as an academic pursuit in the United States.

Professor Benjamin Silliman died in November 1864. He is buried at Grove Street Cemetery in New Haven, Connecticut. Today, only about forty pounds of the Weston meteorite remain, most of which is housed in various museums, including Yale's Peabody Museum and the Smithsonian Institution's National Museum of Natural History.

Benjamin Silliman pioneered the sciences of geology, mineralogy, and paleontology in the United States. He became a leading advocate of science in early nineteenth-century America and inspired many people, from ordinary citizens to scholars, to explore their surroundings. The geologists he trained had a profound influence on the world's understanding of geology. Among those were Amos Eaton, a noted botanist who wrote the first in-depth book about New York State's flora and cofounded Rensselaer Polytechnic Institute in 1824; Edward Hitchcock, a celebrated geologist and paleontologist as well as third president of Amherst College; son-in-law Oliver P. Hubbard; son-in-law James Dwight Dana, a mineralogist, zoologist, and geologist who succeeded Silliman as Professor of Natural History and Chemistry at Yale College; and finally, Benjamin Silliman Jr., a mineralogist who was key in developing the oil industry.

Professor Silliman retired from Yale in 1853 but taught there until 1855. Silliman College at Yale University is named for the professor, as is the mineral Sillimanite.

He kept abreast of scientific ideas without ever silencing his religious faith. It simply never occurred to him the two should conflict; he remained firmly devoted to his faith, without ever hesitating to accept new discoveries.

"Knowledge is nothing but the just and full comprehension of the real nature of things, physical, intellectual, and moral; it is coextensive with the universe of being, both material and spiritual; it reaches back to the dawn of time, and forward to its consummation; nay, it is coeval with eternity, and is inseparable from the incomprehensible existence of

Jehovah. Only one mind, therefore, intuitively embraces the whole," he told his students.[3] More than two hundred years after the meteorite fell, it seems Silliman's sensibilities have come full circle. The rift separating faith from science has, for the most part, closed.

Benjamin Silliman never forgot he was just one generation removed from those who fought the American Revolution. He understood the sacrifices his father made to secure American independence, and he knew his nation needed to become culturally and scientifically independent. In his later years he became a staunch abolitionist. He collected money to arm Northern settlers in their 1856 fight for the Union during what came to be called "Bloody Kansas."

Throughout his tenure at Yale, Silliman never ceased thinking about the immensity of space. He never tired of watching the winking lights of the night sky, regarding stars as the gatekeepers to other planetary systems outside our own. "Nothing can exceed the splendor and magnificence of the starry heavens in a cold and cloudless night," he wrote. "These are probably worlds so distant that our glorious sun has not been seen from them or is discerned only as a twinkling luminous point."[4]

From a time when people planted by moonlight we now understand there is some truth to those superstitions and folklore. But there is scientific foundation. Even so, falling stars—meteorite showers—resonate, and people, even the most logical of them, sometimes wish upon a star.

Yet for all the genteel beauty in the night sky, the thought of nature's violent tendencies never strayed from Silliman's mind. During lectures, he often asked his audience whether meteorites might not one day bring unimagined catastrophe. "May they not one day come down entirely? Shall we desire it! They might sweep away cities and mountains—deeply scar Earth and rear from their own ruins colossal monuments of the great catastrophe."[5]

As such, Benjamin Silliman's study of rocks and gems captured the American imagination and helped fuel today's space program. That scientific literacy and public understanding of science took root is a testament to Benjamin Silliman. For Silliman understood that communicating science removed doubt and superstition. He used science to help unlock the secrets of the universe. In the last half century, the race to the moon influenced a generation of scientists, engineers, geologists, and astronauts. From launching the first satellites to exploring the surface of Mars, science continues to speak to civic pride. Today, scientists explore with their telescopes beyond the Milky Way and try to re-create the big bang in the Large Hadron Collider at the CERN on the border of Switzerland and France.

There have been few times in history when a scientist has excited the public to the extent that Benjamin Silliman did. He captured the American imagination at a time when it was ripe for the taking. He connected with people and dared them to look beyond their borders—both terrestrial and extraterrestrial. He spoke to people during a time of national transition, from a rural nation to a world power. Indeed, America could have contented itself with agriculture and wallowed in peace after the guns of the Revolution quieted. Instead it set its sight on becoming the best, in science, in culture, in military. And Professor Benjamin Silliman was a part of that—because he started looking into Earth's birth one cold December day when the nation was still young.

ACKNOWLEDGMENTS

I would like to thank the staff at Yale University's Sterling Memorial Library as well as the staff at the Historical Society of Philadelphia, the Fairfield Historical Society, the Weston Historical Society, and the Easton Historical Society.

Thank you to Dan Cruson, a historian in Newtown, for answering every question, no matter how seemingly trivial, and helping animate the contours of nineteenth-century life in Weston. Frank Pagliaro of the Easton Historical Society thoughtfully answered many of my inquiries, as did Monty Robson of New Milford.

In addition, Dr. Karl Turekian, Chairman of Geology, Sterling Professor of Geology and Geophysics at Yale University; Barbara Narendra, museum assistant, archivist at Yale's Peabody Museum of Natural History; and Dr. Richard Binzel, Chairman of Planetary Sciences at the Massachusetts Institute of Technology. Their insight proved invaluable.

Thanks also to Dr. Leo O'Connor, chairman of American Studies at Fairfield University, for his advice and feedback; Jill Swenson, an author's advocate in the most literal sense of the word; and Steven L. Mitchell, editor at Prometheus Books, for his excellent feedback.

My parents, to whom this book is dedicated, you gave me both roots and wings.

My children, Nathan and Zoë, your passionate curiosity eternally amazes me—you are my twinkling luminous points.

And, as always, to Pierre; to paraphrase Antoine de Saint-Exupery's *Flight to Arras, 1942*: you are the knot into which my soul is tied.

TIMELINE

1779	Benjamin Silliman born on August 8.
1783	Treaty of Paris signed.
1789	French Revolution begins.
1792	Benjamin Silliman enters Yale.
1795	Timothy Dwight becomes president of Yale.
1796	Silliman graduates Yale with AB.
1798–1799	Silliman apprentices at law firm of Simeon Baldwin.
1799	Silliman graduates Yale with law degree.
1799–1802	Silliman works as tutor at Yale.
1800	Thomas Jefferson elected president.
1800	John Chapman, AKA Johnny Appleseed, begins distributing apple seeds and seedlings to settlers in Ohio.
1801–1805	First Barbary Coast Wars.
1802	United States Military Academy of West Point established on July 4.
1802	Silliman passes bar exam.
1802	Silliman becomes Yale's first professor of chemistry on September 7.
1802	Silliman travels to Philadelphia to learn chemistry.
1803	Louisiana Purchase.
November 5, 1803– March 5, 1804	Silliman returns to Philadelphia.
1804	Silliman delivers first lecture in chemistry at Yale.
1804	Lewis and Clark Expedition.
1805	Silliman leaves for Europe to study.
1806	Noah Webster issues his *Compendious Dictionary of the English Language*.

December 14, 1807	Meteorite falls on Weston, Connecticut.
December 14, 1807	Embargo Act.
1807	Robert Fulton and Robert Livingston build first commercial steamboat.
December 29, 1807	Silliman publishes account of meteorite in *Connecticut Herald.*
1808	James N. Barker, *The Indian Princess, or La Belle Sauvage*; first play having Native American life (that of Pocahontas) as its subject.
March 1808	Silliman's account of the meteorite read before American Philosophical Society.
1808	Silliman begins to lecture publicly.
September 17, 1809	Silliman marries Harriet Trumbull, daughter of Connecticut governor John Trumbull Jr.
September 1810	Silliman reads *Sketch of the Mineralogy of the Town of New Haven* before Connecticut Academy of Arts and Sciences.
1812–1815	War of 1812.
1813	Silliman establishes Yale Medical School.
1814	Embargo Act officially repealed.
1814	Francis Scott Key writes "The Star-Spangled Banner."
December 15, 1814– Janurary 4, 1815	Hartford Convention members talk about secession.
1815	Napoleon defeated and Treaty of Vienna signed.
July 1818	Silliman issues first *American Journal of Science and Arts.*
July 4, 1826	John Adams and Thomas Jefferson die.
1826	First American railroad completed in Quincy, Massachusetts.
1828	John James Audubon publishes first volume of *Birds in America.*
1828	Andrew Jackson elected president.
1846	Mexican War.
1851	The mineral Sillimanite named after Benjamin Silliman.

1851	Silliman's second marriage, to Sarah Isabella McClellan Webb.
1851	Silliman returns to Europe.
1854	Silliman becomes first person to fractionate petroleum by distillation.
1861	Abraham Lincoln elected president.
1861–1865	Civil War.
1863	Emancipation Proclamation.
November 24, 1864	Silliman dies in New Haven, Connecticut, and is buried at Grove Street Cemetery in New Haven.

NOTES

CHAPTER 1

1. Nathaniel Bowditch, "An Estimate of the Height, Direction, Velocity, and Magnitude of the Meteor That Exploded over Weston, in Connecticut, December 14, 1807, With Methods of Calculating Observations Made on Such Bodies," *Memoirs of the American Academy of Arts and Sciences* 3, pt. 2 (1815): 213–36.

2. Ibid.

3. A machine to comb cotton, wool, and other fibers prior to spinning.

4. The process of cleansing, thickening, and shrinking cloth by water.

5. Jack Larkin, *The Reshaping of Everyday Life: 1790–1840* (New York: Harper Perennial, 1988), p. 18.

6. Elizabeth Banks McRury, *More about the Hill: Greenfield Hill* (Youngstown, OH: City Printing Co., 1968), p. 83.

7. Randolph S. Klein, ed., *Science and Society in Early America: Essays in Honor of Whitfield J. Bell Jr.* (Philadelphia: American Philosophical Society, 1986), p. 111.

8. Bowditch, "An Estimate of the Height," p. 227.

9. Ibid.

10. Samuel G. Goodrich, *Peter Parley's Own Story: From the Personal Narrative of the Late Samuel G. Goodrich (Peter Parley)* (New York: Sheldon & Co., 1864), 1:97.

11. Ibid.

12. Ibid., p. 98.

13. Ibid.

14. Ibid., pp. 98–99.

15. Daniel Cruson, historian, Newtown, CT, in discussion with author, December 14, 2007.

16. Steven Shapin, *The Scientific Revolution* (Chicago: University of Chicago Press, 1996), p. 42.

17. Frank Luther Mott, *American Journalism: A History: 1690–1960*, 3rd ed. (New York: Macmillan, 1962), pp. 52–53.

18. Larkin, *The Reshaping of Everyday Life*, p. xiii.

19. Bowditch, "An Estimate of the Height," p. 225.

20. Ibid.

21. Ibid., p. 226.

22. Ibid., p. 227.

23. Goodrich, *Peter Parley's Own Story*, p. 46.

24. Ibid., p. 30.

25. Bowditch, "An Estimate of the Height," p. 214.

26. *The American Journal of Science and Arts*, Second Series. Reprinted from *Memoirs of the Connecticut Academy of Arts and Sciences* 1, no. 1 (1810).

27. Ibid.

28. Jane C. Nylander, *Our Own Snug Fireside: Images of the New England Home 1760–1860* (New Haven, CT: Yale University Press, 1994), p. 6.

29. Ibid., p. 103.

30. Ibid., pp. 20, 23.

31. David H. Kelly and Eugene F. Milone, *Exploring Ancient Skies: An Encyclopedic Survey of Archaeoastronomy* (New York: Springer Science Media Inc., 2005), p. 2.

32. William Henry, *Notes to the American Edition of Henry's Chemistry in an Epitome of Experimental Chemistry*, 2nd ed. from the 5th English ed. (Boston: Institute of Early American History and Culture, 1789, 1810; Chapel Hill: University of North Carolina Press, 1956), p. 17.

33. A. Bevan and J. De Laeter, *Meteorites* (Washington, DC: Smithsonian Institution Press, 2002), p. 30.

34. Frank Pagliaro, Easton Historical Society, e-mail with author, June 4, 2009.

35. Frank Pagliaro, Easton Historical Society, e-mail with author, May 2, 2009.

36. Harry Y. McSween Jr., *Meteorites and Their Parent Planets*, 2nd ed. (Cambridge: Cambridge University Press, 1999), p. xi.

CHAPTER 2

1. Benjamin Silliman, letter to Gold Silliman, December 5, 1807, Silliman Family Papers, Yale University Sterling Memorial Library.

2. Ibid.

3. Ibid.

4. Silliman Family Papers, Yale University Sterling Memorial Library.

5. Jeremiah Day, "A View of the Theories Which Have Been Proposed to

Explain the Origin of Meteoric Stones," *Memoirs of the Academy of Arts and Sciences* 1 (1810): 163–74.

6. Benjamin Silliman, letter, Silliman Family Papers, Yale University Sterling Memorial Library.

7. Nylander, *Our Own Snug Fireside*, p. 6.

8. Thomas J. Farnham, *Weston: The Forging of a Connecticut Town* (Canaan, NH: Phoenix Publishing, 1979).

9. Frank Mott, *American Journalism: A History: 1690–1960*, 3rd ed. (New York: Macmillan, 1962), p. 3.

10. *Connecticut Herald*, December 29, 1807.

11. *Connecticut Journal* published the story on December 24, 1807.

12. David L. Ferror, "Promoting Science through America's Colonial Press," *Early American Review* 11, no. 1 (Summer 1997), http://www.early american.com/revew/summer97/science.html (accessed July 2008).

13. Ibid. The *Gazette* occasionally published articles about Benjamin Franklin's discoveries, which helped readers see science as something reason based, not superstitious.

14. Eric Burns, *Infamous Scribblers: The Founding Fathers and the Rowdy Beginnings of American Journalism* (New York: Public Affairs, 2006), p. 238.

15. Benjamin Silliman, as quoted in John F. Fulton and Elizabeth H. Thomson, *Benjamin Silliman: Pathfinder in American Science* (New York: Henry Schuman, 1947), p. 69.

16. Francis Bacon, as quoted in Steven Shapin, *The Scientific Revolution* (Chicago: University of Chicago Press, 1996), p. 87.

17. Ibid.

18. Ibid., p. 88.

19. Frank Pagliaro, Easton Historical Society, in discussion with author, May 4, 2009.

20. Richard O. Norton, *Rocks from Space: Meteorites and Meteorite Hunters* (Missoula, MT: Mountain Press, 1998), p. 51.

21. Robert T. Dodd, *Thunderstones and Shooting Stars: The Meaning of Meteorites* (Cambridge, MA: Harvard University Press, 1986), p. 4.

22. Frank Pagliaro, Easton Historical Society, in discussion with author, May 6, 2009.

23. *American Journal of Science and Arts*, Second Series, reprinted from *Memoirs of the Connecticut Academy of Arts and Sciences* 1, no. 1 (1810): 328–29.

24. Ibid., p. 326.

25. Ibid.

26. Ibid., p. 329.

27. Ibid.

28. Chandos Michael Brown, *Benjamin Silliman: A Life in the Young Republic* (Princeton, NJ: Princeton University Press, 1989), p. 222.

29. Ibid.

30. Benjamin Silliman and James Kingsley, "An Account of the Meteor," *Memoirs of Connecticut Academy of Arts and Sciences* 1, no. 1 (1810): 149–50.

31. Dirk J. Struik, *Yankee Science in the Making: Science and Engineering in New England from Colonial Times to the Civil War* (New York: Dover, 1991), p. 336.

32. Mary Silliman, letter to Benjamin Silliman, December 1807, Silliman Family Papers, Yale University Sterling Memorial Library.

33. Benjamin Silliman, letter to Mary Silliman, December 27, 1807, Silliman Family Papers, Yale University Sterling Memorial Library; Jeremiah Day, letter to Thomas Day, December 29, 1807, Day Family Papers, Yale University Sterling Memorial Library.

34. Ibid.

35. Jeremiah Day, letter to Thomas Day, Day Family Papers, Yale University Sterling Memorial Library.

36. Mary Silliman, letter to Benjamin Silliman, January 1, 1808, Silliman Family Papers, Yale University Sterling Memorial Library.

37. Ibid.

38. D. T. King Jr. and L. W. Petruny, *The Weston Meteorite 1807, Impact Sites in Fairfield County, Connecticut* (Auburn, AL: Auburn University, 2008).

39. Isaac Bronson, letter to Benjamin Silliman, December 31, 1807, Silliman Family Papers, Yale University Sterling Memorial Library.

40. Jeremiah Day, "A View of the Theories Which Have Been Proposed to Explain the Origin of Meteoric Stones," *Memoirs of the Academy of Arts and Sciences* 1 (1810).

CHAPTER 3

1. George Fischer, *Life of Benjamin Silliman, M.D., LL.D. Late Professor of Chemistry, Mineralogy, and Geology in Yale College. Chiefly from His Manuscript Reminiscences, Diaries, and Correspondence*, 2 vols. (New York: Charles Scribner, 1866), 1:175.

2. Ibid.

3. Richard D. Brown, *Knowledge Is Power: The Diffusion of Information in Early America, 1700–1865* (New York: Oxford University Press, 1989).

4. Mary Silliman, letter to Benjamin Silliman, November 20, 1802, Silliman Family Papers, Yale University Sterling Memorial Library.

5. Ibid.

6. Benjamin Silliman, *Oration before the Society of the Cincinnati for Connecticut*, July 6, 1802, 4th of July Orations, Ella Strong Denison Library, Scripps College, Claremont, CA.

7. John Adams, letter to Abigail Adams, May 12, 1780, Adams Family Correspondence, http://www.masshist.org/adams/quotes.cfm (accessed March 2, 2010).

8. Fischer, *Life of Benjamin Silliman*, pp. 25–26.

9. Timothy Dwight, *Travels in New England and New York*, 4 vols. (New Haven, CT: W. Baynes & Son, 1821–1822), p. 19.

10. Fischer, *Life of Benjamin Silliman*, p. 92.

11. Dirk J. Struik, *Yankee Science in the Making: Science and Engineering in New England from Colonial Times to the Civil War* (New York: Dover, 1991), p. 202.

12. Kelley Brooks Mather, *Yale: A History* (New Haven, CT: Yale University Press, 1999), p. 129.

13. Benjamin Silliman, letters to Mary Silliman, November 18 and 30, 1798, Silliman Family Papers, Yale University Sterling Memorial Library.

14. Dwight as quoted in John F. Fulton and Elizabeth H. Thomson, *Benjamin Silliman: Pathfinder in American Science* (New York: Henry Schuman, 1947), p. 24.

15. Robert V. Bruce, *The Launching of Modern American Science: 1846–1876* (New York: Knopf, 1987).

16. Benjamin Silliman quoted in Fischer, *Life of Benjamin Silliman*, p. 93.

17. Silliman, *Oration before the Society of the Cincinnati for Connecticut.*

18. Mary Silliman, letter to Benjamin Silliman, October 4, 1802, Silliman Family Papers, Yale University Sterling Memorial Library.

19. Jack Larkin, *The Reshaping of Everyday Life: 1790–1840* (New York: Harper Perennial, 1988), pp. 205, 213.

20. Ibid., p. 2.

21. Frances Trollope, *Domestic Manners of the Americans* (New York: Knopf, 1832), p. 278.

22. Simon Baatz, "Squinting at Silliman: Scientific Periodicals in the Early American Republic, 1810–1833," *Isis* 82, no. 2 (June 1991): 223–44.

23. Benjamin Silliman as quoted in Fischer, *Life of Benjamin Silliman*, p. 215, and *American Journal of Science* 3 (1865).

24. Edgar Fahs Smith, *The Life of Robert Hare: An American Chemist, 1781–1858* (Philadelphia: J. B. Lippincott, 1917), p. 7.

25. Ibid.

26. Benjamin Silliman, reminiscences as quoted in Fisher, *Life of Benjamin Silliman*, pp. 100–101.

27. Smith, *The Life of Robert Hare*, p. 8.

28. Linda K. Kerber, *Federalists in Dissent: Imagery and Ideology in Jeffersonian America* (Ithaca and London: Cornell University Press, 1980), p. 83.

29. Benjamin Silliman as quoted in Fulton, *Benjamin Silliman*, p. 36.

30. MacLean as quoted in Ernest Child, *The Tools of the Chemist: Their Ancestry and American Evolution* (New York: Reinhold, 1940), p. 22.

31. Ibid., p. 23.

32. Struik, *Yankee Science in the Making*, p. 204.

33. Benjamin Silliman, *A Journal of Travels in England, Holland and Scotland, and of Two Passages over the Atlantic, in the Years 1805 and 1806; with Considerable Additions, Principally from the Original Manuscripts of the Author* (New Haven, CT: S. Converse, 1820).

34. Benjamin Silliman, journal entry, Friday, March 22, 1805, Silliman Family Papers, Yale University Sterling Memorial Library.

35. Silliman, *A Journal of Travels*, p. 22. Silliman's journal was his first published work. He wrote it at a time when very few Americans went abroad, and of those who did, no one really wrote about it. So his book had the distinction of being the first account by an educated American of impressions of Europe since the American independence.

36. Ibid.

37. Silliman Family Papers, Yale University Sterling Memorial Library.

38. Benjamin Silliman, letter to Mary Silliman, May 3, 1805, Silliman Family Papers, Yale University Sterling Memorial Library.

39. Mary Silliman, letter to Benjamin Silliman, June 7, 1805, Silliman Family Papers, Yale University Sterling Memorial Library.

40. Silliman, *A Journal of Travels*, p. 46.

41. *American Journal of Science* 39, no. 115, second series (January 1865): 8–9. Ironically, Silliman's family had owned slaves under statutory slave law in Connecticut, which was meant to phase out slavery. All slaves born on or after 1784 would be freed at the age of twenty-five. All those born before 1784 were slaves for life.

42. Benjamin Silliman, letter to Mary Noyes, Edinburgh, February 27, 1806, Silliman Family Papers, Yale University Sterling Memorial Library.

43. James Kingsley, letter to Benjamin Silliman, June 29, 1806, Silliman Family Papers, Yale University Memorial Sterling Library.

44. Benjamin Silliman, letter to George Fischer, as quoted in Fischer, *Life of Benjamin Silliman*, pp. 195–96.

CHAPTER 4

1. Benjamin Silliman, *An Introductory Lecture, Delivered in the Laboratory of Yale College* (October 1826), pp. 28–29.

2. http://nationalhumanitiescenter.org/pds/becomingamer/ideas/text4/yale regulations.pdf (accessed February 2010).

3. Ibid.

4. Kelley Brooks Mather, *Yale: A History* (New Haven, CT: Yale University Press, 1999), p. 127.

5. Ernest Child, *The Tools of the Chemist: Their Ancestry and American Evolution* (New York: Reinhold, 1940), p. 6.

6. Johannes Kepler as quoted in Steven Shapin, *The Scientific Revolution* (Chicago: University of Chicago Press, 1996), p. 33.

7. Francis Bacon as quoted in P. R. Masani, "Three Modern Enemies of Science: Materialism, Existentialism, Constructivism," *Kybermetes* 30, no. 3 (2001): 278–94.

8. William Henry, *Notes to the American Edition of Henry's Chemistry in an Epitome of Experimental Chemistry*, 2nd ed. from the 5th English ed. (Boston: Institute of Early American History and Culture, 1789, 1810; Chapel Hill: University of North Carolina Press, 1956), p. 95.

9. Ibid.

10. Ibid.

11. Silliman, *An Introductory Lecture*.

12. George Fischer, *Life of Benjamin Silliman, M.D., LL.D. Late Professor of Chemistry, Mineralogy, and Geology in Yale College. Chiefly from His Manuscript Reminiscences, Diaries, and Correspondence*, 2 vols. (New York: Charles Scribner, 1866), p. 126.

13. Ibid.

14. Ibid.

15. Ibid., p. 89.

16. John C. Greene, *American Science in the Age of Jefferson* (Ames: Iowa State University Press, 1984), p. 21.

17. http://religionanddemocracy.lib.virginia.edu/Jefferson/quotations/jeff 0750.html (accessed February 2010).

18. Alexis de Tocqueville, *Democracy in America*, trans. Arthur Goldhammer (New York: Library of America, 2004), p. 516.

19. Greene, *American Science in the Age of Jefferson*, p. 21.

20. Brooke Hindle, *The Pursuit of Science in Revolutionary America, 1735–1789* (New York: Norton, 1974), p. 382.

21. Silvio Bedini, *Thomas Jefferson: Statesman of Science* (New York: Macmillan, 1990), p. 197.

22. http://www.fordham.edu/halsall/mod/lect/mod14.html (accessed October 2009).

23. John Adams, letter to Mercy Warren, December 25, 1787, in Adams Letters, vol. 2, Massachusetts Historical Society Collections, quoted in Hindle, *The Pursuit of Science in Revolutionary America*, p. 329.

24. Greene, *American Science in the Age of Jefferson*, p. xiii.

25. Hindle, *The Pursuit of Science in Revolutionary America*, p. 6.

26. Benjamin Franklin as quoted in Walter Isaacson, *Benjamin Franklin: An American Life* (New York: Simon & Schuster, 2003), p. 122.

27. Ibid.

28. Simon Baatz, "Squinting at Silliman: Scientific Periodicals in the Early American Republic, 1810–1833," *Isis* 82, no. 2 (June 1991): 225.

29. Isaacson, *Benjamin Franklin*, p. 102.

30. Nan Goodman, "Banishment, Jurisdiction, and Identity in Seventeenth-Century New England: The Case of Roger Williams," *Early American Studies: An Interdisciplinary Journal* 7, no. 1 (Spring 2009): 109–39.

31. Thomas Jefferson, letter to Dr. Thomas Ewell, Monticello, August 1805, as quoted in Thomas Ewell, *Plain Discourses on the Laws or Properties of Matter* (New York: Brisban & Brannan, 1806), pp. 8–9.

32. Benjamin Silliman, *An Introductory Lecture, Delivered in the Laboratory of Yale College* (October 1826), pp. 28–29, Silliman Family Papers, Yale University Sterling Memorial Library.

33. David Hume, *Treatise of Human Nature* (1739–40; London: Penguin, 1985), p. 46.

34. Thomas Jefferson to Isaac McPherson, 1813, as quoted in Thomas O. Jewett, "Thomas Jefferson: Father of Invention," *Early American Review* 3, no. 1 (Winter 2000), http://www.earlyamerican.com/2000_fall/jefferson_paleon.html (accessed August 2009).

35. Henry F. May, *Enlightenment in America* (London: Oxford University Press, 1976), p. 35.

36. De Tocqueville, *Democracy in America*, p. 562.

37. Silliman, *An Introductory Lecture*, pp. 22–23.

CHAPTER 5

1. Barbara B. Oberg, ed., *The Papers of Thomas Jefferson: Volume 30: 1 January 1798 to 31 January 1799 by Thomas Jefferson* (Princeton, NJ: Princeton University Press, 2003), p. 99.

2. Doron Ben-Atar and Barbara Oberg, *Federalists Reconsidered* (Charlottesville: University of Virginia Press, 1998), p. 199.

3. Oberg, *The Papers of Thomas Jefferson*, pp. 381–84.

4. John R. Fitzmier, *New England's Moral Legislator: Timothy Dwight 1752–1817, Religion in North America* (Bloomington: Indiana University Press: 1998), p. 60.

5. Benjamin Silliman, *A Sketch of the Life and Character of President Dwight: Delivered as an Eulogium, in New Haven, February 12, 1817, Before the Academic Body, of Yale College, Composed of the Senatus Academicus, Faculty and Students* (New Haven, CT: Maltby, Goldsmith, 1817).

6. Timothy Dwight, sermon, "A Duty of Americans, at the Present Crisis," July 4, 1800.

7. Thomas Jefferson to Thomas Cooper, November 29, 1802, http://www.etext.virginia.edu.jefferson/quoteations/jeff3.htm (accessed March 11, 2010).

8. Linda K. Kerber, *Federalists in Dissent: Imagery and Ideology in Jeffersonian America* (Ithaca, NY: Cornell University Press, 1980), p. xi.

9. Thomas Jefferson to Benjamin Waring, 1801, http://www.etext.virginia.edu.jefferson/quoteations/jeff3.htm (accessed March 11, 2010).

10. "A Christian Federalist to the Voters of Delaware," September 21, 1800, as quoted in Eugene R. Sheridan, *Jefferson and Religion* (Thomas Jefferson Foundation, Monticello Monograph Series, 1998), pp. 22–23.

11. Noah Webster, "An Oration Pronounced before the Citizens of New Haven on the Anniversary of the Independence of the United States" (New Haven, CT: T. and S. Green); microfilm available. Woodbridge, CT, Research Publications, Inc., 1986. 1 reel; 35mm. (The Eighteenth Century; reel 1515, no. 38).

12. David Fischer Hackett, *The Revolution of American Conservatism: The Federalist Party in the Era of Jeffersonian Democracy* (New York: Harper & Row, 1932, 1965), p. 169; *Hudson Balance*, November 6, 1802.

13. *Connecticut Courant*, quoted in Paul F. Boller, *Presidential Campaigns from George Washington to George W. Bush* (London: Oxford University Press, 2004), p. 12.

14. William Linn, "Serious Considerations on the Election of a President; Addressed to the Citizens of the United States, New York, 1800," as quoted in Sheridan, *Jefferson and Religion*, p. 22.

15. Ibid., p. 48.

16. Jerry W. Knudson, *Jefferson and the Press: Crucible of Liberty* (Columbia: University of South Carolina Press, 2006), p. 47.

17. Ibid., p. 57.

18. James M. Banner Jr., *To the Hartford Convention: The Federalists and the Origins of Party Politics in Massachusetts, 1789–1815* (New York: Knopf, 1970), p. 48.

19. John Adams as quoted in Marable Manning, *Black Leadership* (New York: Columbia University Press, 1999), p. 196.

20. Ralph Henry Gabriel, *Religion and Learning at Yale: The Church of Christ in the College and University, 1757–1957* (New Haven, CT: Yale University Press, 1958), p. 72.

21. Timothy Dwight, *Travels in New England and New York*, 4 vols. (New Haven, CT: W. Baynes & Son, 1821–1822).

22. Timothy Dwight, *Discourse on Some Events of the Last Century in Brick Church in New Haven* (New Haven, CT: Printed by Ezra Read, 1801; Microopaque, Worcester, MA: American Antiquarian Society, 1964).

23. Sheridan, *Jefferson and Religion*, pp. 9, 17.

24. Oberg, *The Papers of Thomas Jefferson*, p. 485.

25. Kerber, *Federalists in Dissent*, p. 83.

26. Benjamin Silliman, "Oration before the Society of the Cincinnati for Connecticut," pp. 6–7.

27. Benjamin Silliman, "An Introductory Lecture, Delivered in the Laboratory of Yale College," October 1826, Silliman Family Papers, Yale University Sterling Memorial Library.

CHAPTER 6

1. Washington Irving, *A Knickerbocker's History of New York* (1809; New York: Biblio Life, 2008), p. 268.

2. Joy Day Buel and Richard Buel, *The Way of Duty: A Woman and Her Family in Revolutionary America* (New York: Norton, 1984), p. xiii.

3. Benjamin Silliman, letter to Mary Silliman, January 10, 1808, Silliman Family Papers, Yale University Sterling Memorial Library.

4. Jeremiah Day, "A View of the Theories Which Have Been Proposed to Explain the Origin of Meteoric Stones," *Memoirs of the Connecticut Academy of Arts and Sciences* 1 (1810): 163–74; David Rittenhouse, *Account of a Meteor, Transactions, American Philosophical Society* 2 (1786).

5. Rittenhouse, *Account of a Meteor*, p. 175.

6. Benjamin Silliman, letter to Joseph Kingsley, Philadelphia, January 23, 1808, Silliman Family Papers, Yale University Sterling Memorial Library.

7. A. Bevan and J. De Laeter, *Meteorites* (Washington, DC: Smithsonian Institution Press, 2002), p. 20.

8. A rock with sharp fragments embedded in clay.

9. Silliman Family Papers, Yale University Sterling Memorial Library.

10. Karl Turekian, Sterling Professor of Geology and Geophysics, Chairman of Geology, Yale University, in discussion with author, November and December 2007.

11. Ibid.

12. Benjamin Silliman, letter to Mary Silliman, January 15, 1808, Silliman Family Papers, Yale University Sterling Memorial Library.

13. *Connecticut Herald*, December 29, 1807.

14. *The Churchmen's Magazine; or, Treasury of Divine and Useful Knowledge*, vol. 4 (New Haven, CT: Oliver Steele, 1807).

15. Ibid.

16. Benjamin Silliman, *The American Journal of Science and Arts* 1, no. 1 (1810), second series, reprinted from *Memoirs of the Connecticut Academy of Arts and Sciences.*

17. Dr. John Brickell, Savannah, Georgia, to Josiah Meigs, February 22, 1809, quoted in *Monthly Anthology and Boston Review* (1809) 6:283.

18. David Judson wrote to Silliman, as quoted in Richard D. Brown. *Knowledge Is Power: The Diffusion of Information in Early America, 1700–1865* (New York: Oxford University Press, 1989), p. 223.

19. James Woodhouse, letter to Benjamin Silliman, January 6 and 8, 1808, Silliman Family Papers, Yale University Sterling Memorial Library.

20. Benjamin Silliman, letter to Joseph L. Kingsley, January 23, 1808, Silliman Family Papers, Yale University Sterling Memorial Library.

21. Benjamin Silliman to James L. Kingsley, January 23, 1808; Benjamin Silliman to James L. Kingsley, January 25, 1808, as quoted in George Fischer, *Life of Benjamin Silliman, M.D., LL.D. Late Professor of Chemistry, Mineralogy, and Geology in Yale College. Chiefly from His Manuscript Reminiscences, Diaries, and Correspondence*, 2 vols. (New York: Charles Scribner and Co., 1866).

22. Benjamin Silliman, letter to Joseph Kingsley, January 23, 1808, Silliman Family Papers, Yale University Sterling Memorial Library.

23. Silliman Family Papers, Yale University Sterling Memorial Library.

24. Silliman Family Papers, Yale University Sterling Memorial Library.

25. Benjamin Silliman, letter to Joseph L. Kingsley, January 23 and 25, 1808, quoted in Fischer, *Life of Benjamin Silliman*, pp. 225–27.

26. Ibid.

27. Jeremiah Day, letter to Thomas Day, February 8, 1809, and Thomas Day to Jeremiah Day, March 10, 1809, Day Family Papers, Yale University Sterling Memorial Library.

28. Frederick Hall, letter to Benjamin Silliman, April 7, 1808, Historical Society Philadelphia.

29. James Woodhouse, letter to Benjamin Silliman, January 6, 1808, Silliman Family Papers, Yale University Sterling Memorial Library.

30. Nathaniel Bowditch, "An Estimate of the Height, Direction, Velocity, and Magnitude of the Meteor that Exploded over Weston, in Connecticut, December 14, 1807. With Methods of Calculating Observations Made on Such Bodies," *Memoirs of the American Academy of Arts and Sciences* 3 (1815): 213–36.

31. Ibid.

32. Day, "A View of the Theories."

33. Ibid.

34. Benjamin Silliman, Silliman Family Papers, Yale University Sterling Memorial Library.

35. Silliman as quoted in John F. Fulton and Elizabeth H. Thomson, *Benjamin Silliman: Pathfinder in American Science* (New York: Henry Schuman, 1947), pp. 138–39.

36. Silliman as quoted in Ralph Henry Gabriel, *Religion and Learning at Yale: The Church of Christ in the College and University, 1757–1957* (New Haven, CT: Yale University Press, 1958), p. 115.

CHAPTER 7

1. Silliman Family Papers, Yale University Sterling Memorial Library.

2. Harry Y. McSween Jr., *Meteorites and Their Parent Planets*, 2nd ed. (Cambridge: Cambridge University Press, 1999), p. 1.

3. O. Richard Norton, *Rocks from Space: Meteorites and Meteorite Hunters* (Missoula, MT: Mountain Press, 1998).

4. Masako M. Shima, S. Murayama, A. Okada, H. Yabuki, and N. Takaoka, "Description, Chemical Composition and Noble Gases of the Chondrite Nogata," *Meteoritics* 18 (June 30, 1983): 87–102.

5. E. Th. Theodossiou, P. G. Niarchos, and V. N. Manimanis, "The Fall of

a Meteorite at Aegos Potamiin *467/6* Be," *Journal of Astronomical History and Heritage* 5, no. 2 (2002): 135–40.

6. William Shakespeare, *Julius Caesar*, act 2, scene 2.

7. Book of Joshua, v. 11–15; Peter Lancaster Brown, *Comets, Meteorites and Men* (New York: Taplinger, 1973), p. 152.

8. Anthony F. Aveni, "Astronomers & Stargazers: Eyeing a Heliocentric Heaven for Planets, Portents & Horoscopes," *Colonial Williamsburg* (Winter 2005–2006): 56–60.

9. Silliman Family Papers, Yale University Sterling Memorial Library.

10. McSween, *Meteorites and Their Parent Planets*, p. 3.

11. Norton, *Rocks from Space*, p. 35.

12. A. Bevan J. De Laeter, *Meteorites* (Washington, DC: Smithsonian Institution Press, 2002), p. 13.

13. Brown, *Comets, Meteorites and Men*, p. 153.

14. McSween, *Meteorites and Their Parent Planets*, p. 2.

15. Silliman Family Papers, Yale University Sterling Memorial Library.

16. Bingley as quoted in McSween, *Meteorites and Their Parent Planets*, p. 3.

17. Norton, *Rocks from Space*, p. 36.

18. Thomas Clap, *Conjectures upon the Nature and Motion of Meteors, Which Are above the Atmosphere* (Norwich, CT: Printed by J. H. Trumbull, 1781), p. 9.

19. Ursula B. Marvin, "Ernst Florens Friedrich Chladni (1756 to 1827)," *Meteoritics* 31 (1996): 545–88.

20. *The Churchmen's Magazine; or, Treasury of Divine and Useful Knowledge*, vol. 4 (New Haven, CT: Oliver Steele, 1807).

21. C. T. Pillinger and J. M. Pillinger, "The Wold Cottage Meteorite: Not Just Any Chondrite," *Meteoritics and Planetary Science* 3 (1996): 589–605.

22. Silliman Family Papers, Yale University Sterling Memorial Library.

23. Ibid.

24. *Churchmen's Magazine*.

25. *New York Times*, as quoted in Neil DeGrasse Tyson, *The Pluto Files: The Rise and Fall of America's Favorite Planet* (New York: Norton, 2009), p. 55.

26. Bevan, *Meteorites*, p. 22.

27. Frank Wlotzka and Fritz Heide, *Meteorites: Messengers from Space*, trans. R. S. Clarke Jr. (1934; Berlin, Germany: Springer-Verlag, 1995), p. 2.

28. Robert T. Dodd, *Thunderstones and Shooting Stars: The Meaning of Meteorites* (Cambridge, MA: Harvard University Press, 1986).

29. Norton, *Rocks from Space*, p. 2; and Steve Connor, "Meteorites Led to Life on Earth," *American Association for the Advancement of Science* (February 21, 1994).

30. Charles S. Cockell, "The Origin and Emergence of Life under Impact Bombardment," *Philosophical Transactions of the Royal Society* (October 29, 2006): 361 (1474); Carl Zimmer, "Meteorites May Have Fostered Life on Earth," *Wired.com* (December 2007), http://www.wired.com/science/planetearth/news/2007/12/dissection_1221 (accessed July 21, 2009).

31. McSween, *Meteorites and Their Parent Planets*, p. 275.

32. Norton, *Rocks from Space*, pp. 15, 48.

CHAPTER 8

1. Sir Arthur Clarke, "Essays on Science and Society: Presidents, Experts and Asteroids," *Science* 280, no. 5369 (June 1998): 1532.

2. Samuel Latham Mitchell, "A Discourse on the Character and Services of Thomas Jefferson, More Especially as a Promoter of Natural and Physical Science," pronounced, by request, before the Lyceum of Natural History, New York, October 11, 1826 (New York: G. & C. Carvill, 1826), p. 36.

3. Ibid.

4. James Truslow Adams, *New England in the Republic 1776–1850* (Boston: Little, Brown, 1926), p. 239.

5. John B. Hoey, "Federal Opposition to the War of 1812," *Early American Review* 3, no. 1 (Winter 2000).

6. Frank Luther Mott, *American Journalism: A History 1690–1960*, 3rd ed. (New York: Macmillan, 1962), pp. 168–69.

7. *National Intelligencer*, December 23, 1807.

8. *New York Evening Post*, February 2, 1808.

9. *National Intelligencer*, December 25, 1807, as quoted in Jerry W. Knudson, *Jefferson and the Press: Crucible of Liberty* (Columbia: University of South Carolina Press, 2006), p. 12.

10. *New York Evening Post*, December 26, 1807.

11. Max Lerner and Robert Schmuhl, *Thomas Jefferson: America's Philosopher-King* (New Brunswick, Canada: Transaction Publishers, 1997), p. 79.

12. Kelley Brooks Mather, *Yale: A History* (New Haven, CT: Yale University Press, 1999), p. 117.

13. Adams, *New England in the Republic 1775–1850*, p. 250.

14. As quoted in Knudson, *Jefferson and the Press*, p. 156.

15. *National Intelligencer*, September 28, 1808, as quoted in Knudson, *Jefferson and the Press*, p. 158.

16. *Port Folio*, July 20, 1808, Boston Repertory.

17. William Cullen Bryant, *The Embargo, or, Sketches of the Times, A Satire* (Boston: printed for the author, 1809).

18. Address of the General Assembly to the People of Connecticut, Hartford, 1809.

19. Stanley Elkins and Eric McKitrick, *The Age of Federalism: The Early American Republic, 1788–1800* (London: Oxford University Press, 1993), p. 381.

20. James M. Banner Jr., *To the Hartford Convention: The Federalists and the Origins of Party Politics in Massachusetts, 1789–1815* (New York: Knopf, 1970), p. 44.

21. Congress repealed the embargo on January 3, 1809, but on March 1, 1809, Congress passed the Non-Intercourse Act. This prohibited French and English ships from entering American ports. It also outlawed American citizens from trading with French or English ships.

22. Barbara B. Oberg, ed., *The Papers of Thomas Jefferson. Volume 30: 1 January 1798 to 31 January 1799 by Thomas Jefferson* (Princeton, NJ: Princeton University Press, 2003).

23. "Science, Exploration, and Travel," http://www.monticello.org/browse/science.html (accessed March 4, 2010).

24. Thomas Jefferson, *A Memoir of the Discovery of Certain Bones of an Unknown Quadruped of the Clawed Kind, in the Western Part of Virginia* (Philadelphia: American Philosophical Society, 1799).

25. Thomas Jefferson to Thomas Cooper, October 7, 1814, as quoted in Andrew Agate Lipscomb, ed., *The Writings of Thomas Jefferson* (Thomas Jefferson Memorial Association of the United States: Washington, DC, 1904), 14:201.

26. Ibid.

27. Henry Raphael, *Thomas Jefferson, Astronomer*, leaflet no. 174, August 1943, *Astronomical Society of the Pacific* 4, p. 184.

28. John P. Foley, *The Jefferson Cyclopedia: A Comprehensive Collection of the Views of Thomas Jefferson* (New York: Funk and Wagnalls, 1900), p. 549.

29. Dr. Richard Binzel, Chairman Planetary Sciences, Massachusetts Institute of Technology, in discussion with the author, December 2007.

30. William Eleroy Curtis, *The True Thomas Jefferson* (Philadelphia: Lippincott, 1901), p. 90.

31. Ibid., p. 359.

32. Catherine Van Cortlandt Mathews, *Andrew Ellicott: His Life and Letters* (New York: Grafton, 1908), pp. 210–11, as quoted in John G. Burke, *Cosmic Debris: Meteorites in History* (Berkeley: University of California Press, 1986), p. 56.

33. Thomas Jefferson to Andrew Ellicott, Lancaster, Pennsylvania, October 6, 1805, and Andrew Ellicott to Thomas Jefferson, October 25, 1805, Jefferson Papers, Library of Congress.

34. Nathaniel Bowditch, "An Estimate of the Height, Direction, Velocity, and Magnitude of the Meteor That Exploded over Weston, in Connecticut, December 14, 1807. With Methods of Calculating Observations Made on Such Bodies," *Memoirs of the American Academy of Arts and Sciences* 3 (1815): 213–36.

CHAPTER 9

1. Silliman Family Papers, Yale University Sterling Memorial Library.

2. Benjamin Silliman, as quoted in George Fischer, *Life of Benjamin Silliman, M.D., LL.D. Late Professor of Chemistry, Mineralogy, and Geology in Yale College. Chiefly from His Manuscript Reminiscences, Diaries, and Correspondence*, 2 vols. (New York: Charles Scribner and Co., 1866), p. 340.

3. Isaac Bronson, letter to Benjamin Silliman, 1808, Silliman Family Papers, Yale University Sterling Memorial Library.

4. Ibid.

5. Darryl Pitt, curator of Macovich Collection New York City, in conversation with author, June 4, 2009.

6. Ibid.

7. John C. Greene, *American Science in the Age of Jefferson* (Ames: Iowa State University Press, 1984), pp. 73–74.

8. Benjamin Silliman, letter written to George Fischer, as quoted in Fischer, *Life of Benjamin Silliman*, p. 214.

9. "Gibbs' Grand Collection of Minerals' Medical Repository Philadelphia," *2d Hexade* 5 (1808): 213–14.

10. Silliman, as quoted in Fischer, *Life of Benjamin Silliman*, p. 219.

11. Benjamin Silliman, *First Principles of Chemistry, For the Use of Colleges and Schools* (Philadelphia: H. C. Peck & T. Bliss, 1854).

12. Ibid.

13. Silliman as quoted in Fischer, *Life of Benjamin Silliman*, p. 259.

14. Timothy Dwight in an address to both Yale College and New Haven citizens in 1825.

15. Benjamin Silliman, "An Introductory Lecture, Delivered in the Laboratory of Yale College," October 1828.

16. Silliman as quoted in Fischer, *Life of Benjamin Silliman*, p. 347.

17. Jane Gregory and Steve Miller, *Science in Public: Communication, Culture, and Curiosity* (New York: Plenum Press, 1998), p. 22.

18. Fischer, *Life of Benjamin Silliman*, p. 258.

19. Benjamin Silliman, "Origin and Progress of Chemistry, Mineralogy and Geology in Yale College with Reminiscences of Personal History," Yale University Sterling Memorial Library, pp. 24–25.

20. Ibid.

21. George H. Daniels, *American Science in the Age of Jackson* (Tuscaloosa: University of Alabama Press, 1968), p. 223.

22. *American Journal of Science* (1865): 5.

23. "Squinting at Silliman: Scientific Periodicals in the Early American Republic, 1810–1833," *Isis* 82, no. 2 (June 1991): 223–44.

24. Horace Hayden, letter to Benjamin Silliman, March 3, 1818, Historical Society Philadelphia.

25. Jacob Bigelow, letter to Benjamin Silliman, March 2, 1818, Historical Society Philadelphia.

26. Henry J. Gratz Family Papers, Historical Society of Pennsylvania.

27. Ibid.

28. Silliman, "An Introductory Lecture," p. 9.

29. *American Journal of Science* (1865): 6.

30. Ibid.

CHAPTER 10

1. Benjamin Silliman, 1817 journal entry, Silliman Family Papers, Yale University Sterling Memorial Library.

2. Leonard G. Wilson, *Benjamin Silliman and His Circle: Studies on the Influence of Benjamin Silliman on Science in America* (New York: Neale Watson Publications, 1979), p. 6.

3. Ibid.

4. Professor James Dwight Dana's Inaugural Discourse as Silliman Professor of Geology in *Yale American Journal of Education*, vol. 1 (1855–1856): 641–42.

5. Wilson, *Benjamin Silliman and His Circle*, p. 115.

6. Ibid., p. 120.

7. Silliman Family Papers, Yale University Sterling Memorial Library.

8. Timothy Dwight, *Memories of Yale Life and Men: 1845–1899* (New York: Dodd Meade,1903), p. 364.

9. Benjamin Silliman Jr., writing to Oliver P. Hubbard, February 2, 1838, Silliman Family Papers, Yale University Sterling Memorial Library.

10. Benjamin Silliman, writing in diary, November 11, 1840, Silliman Family Papers, Yale University Sterling Memorial Library.

11. Brooks Mather Kelley, *Yale: A History* (New Haven, CT: Yale University Press, 1974), p. 246.

12. Ibid., p. 195.

13. Benjamin Silliman, writing to George Fisher as quoted in George Fisher, *Life of Benjamin Silliman, M.D., LL.D. Late Professor of Chemistry, Mineralogy, and Geology in Yale College. Chiefly from His Manuscript Reminiscences, Diaries, and Correspondence. In Two Volumes*, vol. 1 (New York: Charles Scribner, 1866), p. 320.

14. Ibid.

15. Benjamin Silliman, writing to George Fisher, as quoted in ibid., p. 351.

16. Ibid., p. 61.

17. Charles Upham Shepard, *Treatise on Mineralogy* (New Haven, CT: Published by the author, 1857), 1:iii.

18. Charles Upham Shepard, writing to Benjamin Silliman, December 21, 1851, as quoted in Shepard, *Treatise*, p. 96.

19. Russell H. Chittenden, *History of the Sheffield Scientific School of Yale University, 1846–1922*, 2 vols. (New Haven: 1922), 1:28–29.

20. Benjamin Silliman Sr. to Joseph Sheffield, Silliman Family Papers, Yale University Sterling Memorial Library.

21. Benjamin Silliman, writing in journal, August 27, 1836, as quoted in Fisher, *Life of Benjamin Silliman*, pp. 378–79.

CHAPTER 11

1. Ron Seely, "Meteorite Found in Southwestern Wisconsin," *Wisconsin State Journal* (April 16, 2010).

2. Mark Hammergren, astronomer from Adler Planetarium in Chicago, as reported in ibid.

3. Seely, "Meteorite Found in Southwestern Wisconsin."

4. Andrew Jacobs, "Where Prices Are Out of This World," *New York Times*, November 18, 1997, http://www.nytimes.com/1997/11/18/us/where_prices_are _out_of_this_world .htm (accessed March 6, 2009).

5. Harry Y. McSween and D. S. Lauretta, *Meteorites and the Early Solar System* (Tucson: University of Arizona Press, 2006).

6. Harry Y. McSween, *Meteorites and Their Parent Planets*, 2nd ed. (Cambridge: Cambridge University Press, 1999), p. 261.

7. Ibid., p. 254.

8. Ibid., p. 257.

9. John G. Burke, *Cosmic Debris: Meteorites in History* (Berkeley: University of California Press, 1986), p. 312.

10. Ibid., pp. 315–16.

11. Ibid., p. 276.

12. Alex Bevan and John de Laeter, *Meteorites: A Journey through Space and Time* (Washington, DC: Smithsonian Institution Press, 2002), p. 22.

13. Ibid., p. 26.

14. Ibid., p. 54.

15. Ibid., p. 66.

16. Ibid., p. 151.

17. Don Brownlee, "Stardust's Big Surprise," July 6, 2007, http://www.stardust.jpl.nasa.gov/news/news113 (accessed May 5, 2010).

18. Ibid.

19. "Apophis: The Asteroid That Could Smash into the Earth on Friday, April 13th, 2036," www.deepastronomy.com/apophis-asteroid-could-hit-earth (accessed May 10, 2010).

20. "Predicting Apophis' Earth Encounters in 2029 and 2036," http://neo.jpl.nasa.gov/apophis (accessed May 10, 2010).

21. Andrew Lawler, "What to Do before the Asteroid Strikes: The Doomsday Rock Is Out There. It's Just a Matter of Time . . ." *Discovery Magazine*, November 1, 2007.

22. "Predicting Apophis' Earth Encounters in 2029 and 2036."

23. Ibid.

24. "Apophis: The Asteroid That Could Smash into the Earth on Friday, April 13th, 2036."

25. Benjamin Silliman, as quoted in Burke, *Cosmic Debris*, p. 66.

26. "Predicting Apophis' Earth Encounters in 2029 and 2036."

27. Ibid.

EPILOGUE

1. Edmund Turney, "The Meteor," in *Memorial Poems: The Old Schoolhouse; And Other Occasional Pieces* (New York: W. H. Kelley & Bro., 1864).

2. Benjamin Silliman as quoted in Turney, *Memorial Poems*, p. 41.

3. Benjamin Silliman, lab oration, 1826, p. 47.

4. Ibid.

5. Benjamin Silliman, as quoted in John G. Burke, *Cosmic Debris: Meteorites in History* (Berkeley: University of California Press, 1986), p. 66.

BIBLIOGRAPHY

Adams, Henry. *History of the United States: The United States in 1800.* Vol. 1. New York: Charles Scribner's Sons, 1889.

Adams, James Truslow. *New England in the Republic 1776–1850.* Boston: Little, Brown, 1926.

———. *Revolutionary New England, 1691–1776. The History of New England in Three Volumes.* Vol. 2. New York: Cooper Square Publishers, 1968.

Adams, John. Letter to Abigail Adams. May 12, 1780. Adams Family Correspondence. http://www.masshist.org/adams/quotes.cfm (accessed March 2, 2010).

Allen, Thomas A. *The Geological Revolution in American Time.* Richmond, VA: University of Richmond, 2001.

American Journal of Science 39, no. 115. Second series. January 1865.

American Journal of Science and Arts. Second series. Reprinted from *Memoirs of the Connecticut Academy of Arts and Sciences*, Vol. 1, no. 1, 1810.

Aveni, Anthony F. "Astronomers & Stargazers: Eyeing a Heliocentric Heaven for Planets, Portents & Horoscopes." *Colonial Williamsburg* (Winter 2005–2006): 56–60.

Austen, Barbara E., and Barbara D. Bryan. *Images of America: Fairfield, Connecticut.* Dover, NH: Arcadia Publishing, 1997.

Baatz, Simon. *Patronage, Science, and Ideology in an American City: Patrician Philadelphia 1800–1860.* Fiftieth anniversary ed. Philadelphia: University of Pennsylvania, 1986.

———. "Squinting at Silliman: Scientific Periodicals in the Early American Republic, 1810–1833." *Isis* 82, no. 2 (June 1991): 223–44.

Banner, James M., Jr. *To the Hartford Convention: The Federalists and the Origins of Party Politics in Massachusetts, 1789–1815.* New York: Knopf, 1970.

Bedini, Silvio A. *Thomas Jefferson: Statesman of Science.* New York: Macmillan, 1990.

Ben-Atar, Doron, and Barbara B. Oberg. *Federalists Reconsidered.* Charlottesville: University of Virginia Press, 1998.

Bevan, A., and J. De Laeter. *Meteorites.* Washington, DC: Smithsonian Institution Press, 2002.

Binzel, Dr. Richard, chairman, Planetary Sciences, Massachusetts Institute of Technology, in discussion with the author, December 2007.

Boller, Paul F. *Presidential Campaigns from George Washington to George W. Bush.* Oxford: Oxford University Press, 2004.

Bowditch, Nathaniel. "An Estimate of the Height, Direction, Velocity, and Magnitude of the Meteor That Exploded over Weston, in Connecticut, December 14, 1807. With Methods of Calculating Observations Made on Such Bodies." *Memoirs of the American Academy of Arts and Sciences* 3 (1815): 213–36.

Brown, Chandos Michael. *Benjamin Silliman: A Life in the Young Republic.* Princeton, NJ: Princeton University Press, 1989.

Brown, Peter Lancaster. *Comets, Meteorites and Men.* New York: Taplinger Publishing, 1973.

Brown, Richard D. *Knowledge Is Power: The Diffusion of Information in Early America, 1700–1865.* New York: Oxford University Press, 1989.

Bruce, Robert V. *The Launching of Modern American Science: 1846–1876.* New York: Knopf, 1987.

Bryant, William Cullen. *The Embargo, Or, Sketches of the Times, A Satire.* Boston: Printed for the author, 1809.

Buckley, William F., Jr. *God & Man at Yale: The Superstitions of Academic Freedom.* Washington, DC: Regnery, 1986.

Buel, Joy Day, and Richard Buel. *The Way of Duty: A Woman and Her Family in Revolutionary America.* New York: Norton, 1984.

Burke, John G. *Cosmic Debris: Meteorites in History.* Berkeley: University of California Press, 1986.

Burns, Eric. *Infamous Scribblers: The Founding Fathers and the Rowdy Beginnings of American Journalism.* New York: Public Affairs, 2006.

Caswell, Alexis. "Memoir of Benjamin Silliman, Sr. 1779–1864." Read before the National Academy, January 25, 1866. *Biographical Memoirs.* National Academy of Sciences, Vol. 9. Washington, DC: National Academy of Sciences.

Cerniglia, Keith A. "The American Almanac and the Astrology Factor." *Early American Review* (Winter/Spring 2003). http://www.earlyamerica.com/review/2003_winter_spring/almanac.htm (accessed March 6, 2009).

Child, Ernest. *The Tools of the Chemist: Their Ancestry and American Evolution.* New York: Reinhold, 1940.

Chittenden, Russell H. *History of the Sheffield Scientific School of Yale University, 1846–1922.* 2 vols. New Haven, CT: Yale University Press, 1922.

The Churchmen's Magazine; or, Treasury of Divine and Useful Knowledge. Vol. 4. New Haven, CT: Oliver Steele & Co., 1807.

Clap, Thomas. *Conjectures upon the Nature and Motion of Meteors, Which Are above the Atmosphere.* Norwich, CT: Printed by J. H. Trumbull, 1781.

Clark, Austin H. "Thomas Jefferson as Scientist." *Journal of the Washington Academy of Sciences* 33, no. 7 (1943).

Clarke, Arthur Sir. "Essays on Science and Society: Presidents, Experts and Asteroids." *Science* 280, no. 5369 (June 1998): 1532.

Cockell, Charles S. "The Origin and Emergence of Life under Impact Bombardment." *Philosophical Transactions of the Royal Society* 361, no. 1474 (October 29, 2006).

Connecticut Herald, 1807.

Connecticut Historical Records Survey Project, Division of Professional and Service Projects, Works Projects Administration. "Inventory of the Town and City Archives of Connecticut." Vol. 21, no. 1. Weston, CT: May 1940.

Connecticut Journal, December 24, 1807.

Connor, Steve. "Meteorites Led to Life on Earth." American Association for the Advancement of Science. February 21, 1994.

Cruson, Daniel, historian, Newtown, CT, in discussion with the author, December 14, 2007, February 2008, e-mail message to author.

Cummin, Katharine Hewitt. *Connecticut Militia General: Gold Selleck Silliman.* American Revolution Bicentennial Commission of Connecticut. Hartford, CT: Knopf, 1979, 1987.

Curtis, William Eleroy. *The True Thomas Jefferson.* Philadelphia: J. W. Lippincott, 1901.

Dall'olmo, U. "Meteors, Meteor Showers and the Middle Ages." *Journal for the History of Astronomy* 9 (1978): 123–34.

Daniels, George H. *American Science in the Age of Jackson.* Tuscaloosa: University of Alabama Press, 1968.

———., ed. *Nineteenth-Century American Science: A Reappraisal.* Chicago: Northwestern University Press, 1972.

Day, Jeremiah. "A View of the Theories Which Have Been Proposed to Explain the Origin of Meteoric Stones." *Memoirs of the Connecticut Academy of Arts and Sciences.* 1810 (1):163–74.

Day Family Papers, Yale University Sterling Memorial Library.

DeGrasse Tyson, Neil. *The Pluto Files: The Rise and Fall of America's Favorite Planet.* New York: Norton, 2009.

De Tocqueville, Alexis. *Democracy in America.* Trans. by Arthur Goldhammer. New York: Library of America, 2004.

Dodd, Robert T. *Thunderstones and Shooting Stars: The Meaning of Meteorites.* Cambridge, MA: Harvard University Press, 1986.

Dwight, Theodore. *History of the Hartford Convention with a Review of the*

Policy of the United States Government, Which Led to the War of 1812. New York: N. and J. White; Boston: Russell, Odiorne, 1833.

Dwight, Timothy. *Discourse on Some Events of the Last Century in Brick Church in New Haven.* New Haven, CT: Printed by Ezra Read, 1801. Microopaque. Worcester, MA: American Antiquarian Society, 1964.

———. "A Duty of Americans, at the Present Crisis." Sermon. July 4, 1800.

———. *Memories of Yale Life and Men: 1845–1899.* New York: Dodd Meade, 1903.

———. *Travels in New England and New York.* 4 vols. New Haven, CT: W. Baynes & Son, 1821–1822.

Elkins, Stanley, and Eric McKitrick. *The Age of Federalism: The Early American Republic, 1788–1800.* Oxford: Oxford University Press, 1993.

Ellicott, Andrew, letter to Thomas Jefferson, Lancaster, PA, October 6, 1805. Jefferson Papers, Library of Congress, October 25, 1805.

Ellis, Joseph J. *American Sphinx: The Character of Thomas Jefferson.* New York: Vintage Books, 1996.

Ewell, Thomas. *Plain Discourses on the Laws or Properties of Matter.* New York: Brisban & Brannan, 1806.

Failla, Kathleen Saluk. *Images of America: Weston.* Charleston, SC: Arcadia, 2003.

Farnham, Thomas J. *Weston: The Forging of a Connecticut Town.* Canaan, NH: Phoenix, 1979.

Farrell, Winslow, former member of JET Propulsion Lab, NASA, in discussion with author, November 27, 2007.

Farrington, O. C. "The Worship and Folk-lore of Meteorites." *Journal of American Folklore* 50 (1913): 199–208.

Fedor, Ferenz, and Lawrence Schwingel. *Connecticut Cradle of Democracy: An Historical Review of the Birth of Our Nation.* England: Pentland Press, 1996.

Ferror, David L. "Promoting Science through America's Colonial Press." *Early American Review* 11, no. 1 (Summer 1997). http://www.earlyamerican.com/revew/summer97/science.html (accessed July 2008).

Fischer, George. *Life of Benjamin Silliman, M.D., LL.D. Late Professor of Chemistry, Mineralogy, and Geology in Yale College. Chiefly from His Manuscript Reminiscences, Diaries, and Correspondence.* 2 vols, vol. 1. New York: Charles Scribner, 1866.

Fitzmier, John R. *New England's Moral Legislator: Timothy Dwight 1752–1817.* Bloomington: Indiana University Press, 1998.

Foley, John P. *The Jefferson Cyclopedia: A Comprehensive Collection of the Views of Thomas Jefferson.* New York: Funk and Wagnalls, 1900.

Fulton, John F., and Elizabeth H. Thomson. *Benjamin Silliman: Pathfinder in American Science.* New York: Henry Schuman, 1947.

Gabriel, Ralph Henry. *Religion and Learning at Yale: The Church of Christ in the College and University, 1757–1957.* New Haven, CT: Yale University Press, 1958.

Gay, Peter. *The Enlightenment: An Interpretation: The Science of Freedom.* New York: Norton, 1969.

"Gibbs's Grand Collection of Minerals' Medical Repository Philadelphia." *2d Hexade* 5 (1808): 213–14.

Goodman, Nan. "Banishment, Jurisdiction, and Identity in Seventeenth-Century New England: The Case of Roger Williams." *Early American Studies: An Interdisciplinary Journal* 7, no. 1 (Spring 2009): 109–39.

Goodrich, Samuel G. *Peter Parley's Own Story: From the Personal Narrative of the Late Samuel G. Goodrich (Peter Parley).* New York: Sheldon, 1864.

Gratz, Henry J. Family Papers. Historical Society of Pennsylvania.

Greene, John C. *American Science in the Age of Jefferson.* Ames: Iowa State University Press, 1984.

———. "Some Aspects of American Astronomy 1750–1815," *Isis* 45, no. 4 (December 1954): 339–58.

Greene, John C., and John G. Burke. "The Science of Minerals in the Age of Jefferson." *Transactions of the American Philosophical Society* 68, no 4. (1978): 1–113.

Gregory, Jane, and Steve Miller. *Science in Public: Communication, Culture, & Credibility.* New York: Plenum Press, 1998.

Hackett, David Fischer. *The Revolution of American Conservatism: The Federalist Party in the Era of Jeffersonian Democracy.* 1932; New York: Harper & Row, 1965.

Hare, Robert. Letter to Benjamin Silliman. 1824. Philadelphia Historical Society.

Henry, William. "The Elements of Experimental Chemistry." *Transactions of the American Philosophical Society*, 9th ed., vol. 2 (1809).

———. *Notes to the American Edition of Henry's Chemistry in an Epitome of Experimental Chemistry.* 2nd ed. from the 5th English ed. Boston, 1810 (1789). Published for the Institute of Early American History and Culture. Chapel Hill: University of North Carolina Press, 1956.

Herbert, Thomas. *Yale Men & Landmarks in Old Connecticut.* New Haven, CT: Yale University Press, 1967.

Hindle, Brooke. *The Pursuit of Science in Revolutionary America 1735–1789.* New York: Norton, 1974.

Hoey, John B. "Federal Opposition to the War of 1812." *Early American Review* 3, no. 1 (Winter 2000). Archiving Early America (accessed August 10, 2009).

Horton, Gerald James. "The Advancement of Science and Its Burdens." In *The Jefferson Lecture and Other Essays*. Cambridge: Cambridge University Press, 1986.

The Hudson Balance, November 6, 1802.

Hume, David. *Treatise of Human Nature*. 1739–1740; London: Penguin Books, 1985.

Hutchison, David. *The Search for Our Beginning: An Enquiry, Based on Meteorite Research, into the Origin of Our Planet and Life*. London: Oxford University Press, 1983.

Irving, Washington. *A Knickerbocker's History of New York*. 1809; New York: Biblio Life, 2008.

Isaacson, Walter. *Benjamin Franklin: An American Life*. New York: Simon & Schuster, 2003.

Jacobs, Andrew. "Where Prices Are Out of This World." *New York Times*, November 18, 1997. http://www.nytimes.com/1997/11/18/us/where_prices _are _out_of_this_world.htm (accessed March 6, 2009).

Jefferson, Thomas. Letter to John Wise, February 12. *American Historical Review* 3 (1798): 488–89.

———. *A Memoir of the Discovery of Certain Bones of an Unknown Quadruped of the Clawed Kind, in the Western Part of Virginia*. American Philosophical Society, 1799.

Jewett, Thomas O. "Thomas Jefferson: Father of Invention." *Early American Review* 3, no. 1 (Winter 2000). Archiving Early America. http://www.early america.com/review/winter2000/jefferson.html (accessed March 6, 2009).

———. "Thomas Jefferson: Paleontologist." *Early American Review* 3, no. 2 (Fall 2000). Archiving Early America. http://www.earlyamerican.com/ 2000_fall/jefferson_paleon.html (accessed August 2009).

Joshua, Book of. Revised Standard Edition. Chapters 11–15.

Keeter, Scott, David Masci, and Gregory Smith. "Science in America: Religious Belief and Public Attitudes," December 18, 2007. Pew Forum on Religion and Public Life, Pew Center, Washington, DC.

Kelly, David H., and Eugene F. Milone. *Exploring Ancient Skies: An Encyclopedic Survey of Archaeoastronomy*. New York: Springer Science Media, 2005.

Kerber, Linda K. *Federalists in Dissent: Imagery and Ideology in Jeffersonian America*. Ithaca, NY: Cornell University Press, 1980.

Kielbowicz, Richard B. *News in the Mail: The Press, Post-Office, and Publishing Information 1200–1860.* New York: Greenwood Press, 1989.

King, D. T., Jr., and L. W. Petruny. *The Weston Meteorite (1807)—Impact Sites in Fairfield Country, Connecticut.* Lunar and Planetary Science. Auburn, AL: Auburn University, 2008.

Kingsley, Joseph. Kingsley Family Letters and Diaries Extracts. 1805. Yale University Sterling Memorial Library.

Klein, Randolph S., ed. *Science and Society in Early America: Essays in Honor of Whitfield J. Bell Jr.* American Philosophical Society. December 1986.

Knudson, Jerry W. *Jefferson and the Press: Crucible of Liberty.* Columbia: University of South Carolina Press, 2006.

Koch, Adrienne, and William Peden, eds. *The Life and Selected Writings of Thomas Jefferson.* Modern Library of New York: Random House, 1944.

Kohlstedt, Sally Gregory, Michael M. Sokal, and Bruce V. Lewenstein. "The Establishment of Science in America: 150 Years of the American Association for the Advancement of Science." 1999.

Kuslan, Louis I. *Connecticut Science, Technology, and Medicine in the Era of the American Revolution.* Hartford: American Revolution Bicentennial Commission of Connecticut, 1978.

Lange, Erwin F. "The Founders of American Meteoritics." *Meteoritics* 10, no. 3 (September 1975).

Larkin, Jack. *The Reshaping of Everyday Life: 1790–1840.* New York: Harper Perennial, 1988.

Larson, Edward J. *A Magnificent Catastrophe: The Tumultuous Election of 1800, America's First Presidential Campaign.* New York: Free Press, 2007.

Lawler, Andrew. "What to Do before the Asteroid Strikes: The Doomsday Rock Is Out There. It's Just a Matter of Time . . ." *Discovery Magazine,* November 2007.

Lerner, Max, and Robert Schmuhl. *Thomas Jefferson: America's Philosopher-King.* New Brunswick: Transaction, 1997.

Lipscomb, Andrew Agate, ed. *The Writings of Thomas Jefferson,* vol. 10. Published under the auspices of the Thomas Jefferson Memorial Association of the United States: Washington, DC, 1904.

Manning, Marable. *Black Leadership.* New York: Columbia University Press, 1999.

Marvin, Ursula B. "Ernst Florens Friedrich Chladni (1756 to 1827)." *Meteoritics* 31 (1996): 545–88.

Masani, P. R. "Three Modern Enemies of Science: Materialism, Existentialism, Constructivism." *Kybermetes* 30, no. 3 (2001): 278–94.

Mather, Kelley Brooks. *Yale: A History*. New Haven, CT: Yale University Press, 1999.

Mathews, Catherine Van Cortlandt. *Andrew Ellicott: His Life and Letters*. New York: Grafton, 1908.

May, Henry F. *Enlightenment in America*. Oxford: Oxford University Press, 1976.

McCurdy, Howard. *Space and the American Imagination*. Washington, DC: Smithsonian, 1999.

McHone, Greg. *Great Day Trips to Discover the Geology of Connecticut*. Wilton, CT: Perry Heights Press, 2004.

McRury, Elizabeth Banks. *More about the Hill: Greenfield Hill*. Youngstown, OH: City Printing, 1968.

McSween, Harry Y., Jr. *Meteorites and Their Parent Planets*. 2nd ed. Cambridge: Cambridge University Press, 1999.

Merwin, George H. *Church & Parish of Greenfield "Ye Church & Parish of Greenfield: The Story of an Historic Church in an Historic Town"* 1725; New Haven, CT: Tuttle Morehouse and Taylor Press, 1913.

Miller, A. M. "Meteorites." *Scientific Monthly* 17 (1923): 435–48.

Mitchell, Samuel Latham. *A Discourse on the Character and Services of Thomas Jefferson, More Especially as a Promoter of Natural and Physical Science*. Pronounced, by request, before the Lyceum of Natural History, New York, October 11, 1826. New York: G. & C. Carvill, 1826.

Monthly Anthology and Boston Review, 1809.

Morais, Herbert M. *Deism in Eighteenth-Century America*. New York: Russell & Russell, 1960.

Morison, Samuel Eliot. *The Intellectual Life of Colonial New England*. Ithaca, NY: Cornell University Press, 1967.

Mott, Frank Luther. *American Journalism: A History 1690–1960*. 3rd ed. New York: Macmillan, 1962.

National Intelligencer, December 23, 1807.

Narendra, Barbara. Archivist, Yale University, Department of Geology, in discussion with author, November 2007, and e-mail message to author.

———. "Benjamin Silliman and the Peabody Museum." *Discovery* 14 (1979): 13–29.

Nelkin, Dorothy. *Selling Science: How the Press Covers Science and Technology*. New York: W. H. Freeman, 1987.

New York Evening Post, December 26, 1807.

New York Evening Post, February 2, 1808.

Norton, O. Richard. *Rocks from Space: Meteorites and Meteorite Hunters.* Missoula, MT: Mountain Press, 1998.

Nylander, Jane C. *Our Own Snug Fireside: Images of the New England Home 1760–1860.* New Haven, CT: Yale University Press, 1994.

Oberg, Barbara B., ed. *The Papers of Thomas Jefferson. Volume 30: 1 January 1798 to 31 January 1799 by Thomas Jefferson.* Princeton, NJ: Princeton University Press, 2003.

Pagliaro, Frank. Easton Historical Society. E-mail messages to author, May 2 and June 4, 2009.

Payne, William W., and Charlotte R. Willard, eds. *Popular Astronomy* 1 (1893–1894). Northfield, MN: Goodsell Observatory of Carleton College, 1894.

Peden, William, ed. *Thomas Jefferson Notes on the State of Virginia.* Chapel Hill: University of North Carolina Press, 1955.

Pillinger, C. T., and J. M. Pillinger. "The Wold Cottage Meteorite: Not Just Any Chondrite." *Meteoritics and Planetary Science* 3 (1996): 589–605.

Pitt, Darryl. Curator, Macovich Collection, New York, in conversation with author, June 4, 2009.

Port Folio, July 20, 1808, Boston Repertory.

Prince, Cathryn J. "Meteor Blazes Back." *Christian Science Monitor*, December 14, 2007, p. 8.

Raphael, Henry. *Thomas Jefferson, Astronomer.* Leaflet no. 174, August 1943. Astronomical Society of the Pacific, vol. 4.

Rashid, Salem. "Political Economy and Geology in the Early Nineteenth Century: Similarities and Contrasts." *History of Political Economy* 13, no. 4 (1981): 726–44.

Reeds, C. A. "Comets, Meteors, and Meteorites." *Natural History* 33, no. 3 (1933): 311–24.

Rittenhouse, David. *Account of a Meteor.* Transactions, American Philosophical Society 2, 1786.

Robson, Monty. Director, Observatory, New Milford, CT, in discussion with the author, December 2007, and e-mail message to the author.

Seely, Ron. "Meteorite Found in Southwestern Wisconsin." *Wisconsin State Journal*, April 16, 2010.

Serviss, Garrett. *Curiosities of the Sky.* N.p.: Sattre Press, 1909.

Shakespeare, William. *Julius Caeser*, act 2, scene 2.

Shapin, Steven. *The Scientific Revolution.* Chicago: University of Chicago Press, 1996.

Sheridan, Eugene R. *Jefferson and Religion.* Thomas Jefferson Foundation. Monticello Monograph Series, 1998.

Shepard, Charles Upham. *Treatise on Mineralogy*. New Haven, CT: Published by the author, 1857.

Shima, Masako M., S. Murayama, A. Okada, H. Yabuki, and N. Takaoka. "Description, Chemical Composition and Noble Gases of the Chondrite Nogata." *Meteoritics* 18 (June 30, 1983): 87–102.

Shoeck, Helmut, ed. *Power of Print in American History, 1776–1976*. New York: Library of Congress, Regis Paper Co., 1977.

Silliman, Benjamin. *First Principles of Chemistry, For the Use of Colleges and Schools*. Philadelphia: H. C. Peck & T. Bliss, 1854.

———. *A Journal of Travels in England, Holland and Scotland, and of Two Passages over the Atlantic, in the Years 1805 and 1806; with Considerable Additions, Principally from the Original Manuscripts of the Author.* New Haven, CT: S. Converse, 1820.

———. "An Introductory Lecture, Delivered in the Laboratory of Yale College." October 1826.

———. "An Introductory Lecture, Delivered in the Laboratory of Yale College." October 1828.

———. "Oration before the Society of the Cincinnati for Connecticut." July 6, 1802. 4th of July Orations. Ella Strong Denison Library, Scripps College, Claremont, CA.

———. *Outline of the Course of Geological Lectures Given in Yale College*. New Haven, CT: Hezekiah Howe, 1829.

———. *A Sketch of the Life and Character of President Dwight: Delivered as an Eulogium, in New Haven, February 12, 1817, before the Academic Body, of Yale College, Composed of the Senatus Academicus, Faculty and Students*. New Haven, CT: Maltby, Goldsmith & Co., 1817.

Silliman, Benjamin, and James Kingsley. "An Account of the Meteor." *American Journal of Science and Arts*. Second Series. Reprinted from *Memoirs of the Connecticut Academy of Arts and Sciences* 1, no. 1 (1810).

Silliman Family Papers. Yale University Sterling Memorial Library.

Slotten, Hugh Richard. *Patronage, Practice and the Culture of American Science: Alexander Dallas Bache & the US Coast Survey*. Cambridge: Cambridge University Press, 1994.

Smith, Edgar Fahs. *James Woodhouse: A Pioneer in Chemistry, 1770–1809*. Philadelphia: John C. Winston, 1918.

———. *The Life of Robert Hare: An American Chemist, 1781–1858*. Philadelphia: J. B. Lippincott, 1917.

Spinelli, Tony. "The Day the Sky Fell." *Connecticut Post*, April 22, 2004.

Streitmatter, Rodger. *Mightier Than the Sword: How the News Media Have Shaped American History.* New York: Westview Press, 1997.

Struik, Dirk J. *Yankee Science in the Making: Science and Engineering in New England from Colonial Times to the Civil War.* New York: Dover, 1991.

Theodossiou, E. Th., P. G. Niarchos, and V. N. Manimanis. "The Fall of a Meteorite at Aegos Potamiin *467/6* Be." *Journal of Astronomical History and Heritage* 5, no. 2 (2002): 135–40.

Thomas, Paul J., Christopher F. Chyba, and Christopher P. McKay. *Comets and the Origin and Evolution of Life.* New York: Springer-Verlag, 1997.

Todd, Timmons. *Science and Technology in Nineteenth-Century America.* Santa Barbara, CA: Greenwood Press, 2005.

Trollope, Frances. *Domestic Manners of the Americans.* New York: Knopf, 1832.

Turekian, Dr. Karl. Chairman of Geology Department, Sterling Professor of Geology & Geophysics, Yale University, in discussion with author, November and December 2007.

Turney, Edmund. "The Meteor." In *Memorial Poems: The Old School-house; And Other Occasional Pieces.* New York: W. H. Kelley & Bro., 1864.

Virgiliu, Pop G. *Who Owns the Moon? Extraterrestrial Aspects of Land and Mineral Resources Ownership.* Vol. 4. New York: Springer Space Regulations Library Series, 2009.

Washington, H. A., ed. *The Writings of Thomas Jefferson.* Letter to Rev. James Madison. Paris, July 19, 1788. Vol. 2 (1853–1854): 431.

Wasson, John T. *Meteorites: Their Record of Early Solar-System History.* New York: W. H. Freeman, 1985.

Webster, Noah. "An Oration Pronounced before the Citizens of New Haven on the Anniversary of the Independence of the United States." New Haven, CT: T. and S. Green. Available microfilm. Woodbridge, CT: Research Publications, 1986. 1 reel; 35mm. (Eighteenth Century; reel 1515, no. 38).

White, Andrew Dickson. *Memorial Addresses at the Unveiling of the Bronze Statue of Professor Benjamin Silliman: At Yale College, June 24, 1884.* New Haven, CT: Tuttle, Morehouse, & Taylor Printers, 1885.

Wlotzka, Frank, and Fritz Heide. *Meteorites: Messengers from Space.* Original 1934. Trans. Dr. R. S. Clarke Jr. Berlin: Springer-Verlag, 1995.

Wilson, Leonard G. *Benjamin Silliman and His Circle: Studies on the Influence of Benjamin Silliman on Science in America.* New York: Science History Publications, 1979.

Wood, John A. *Meteorites and the Origin of Planets.* New York: McGraw-Hill, 1968.

Wyne, James. "Benjamin Silliman." *Harper's New Monthly Magazine* 25, no. 148, September 1862.

Yale University, "Student Regulations at Yale College," http://nationalhumanities center.org/pds/becomingamer/ideas/text4/yaleregulations.pdf (accessed February 2010).

Zimmer, Carl. "Meteorites May Have Fostered Life on Earth." *Wired.com,* http://www.wired.com/science/planetearth/news/2007/12/dissection_1221 (accessed July 21, 2009).

INDEX